PROPERTY OF KAKASHI steal &
you die!

The COMPLETE Guide to
PREHISTORIC LIFE

A FIREFLY BOOK

Published by Firefly Books Ltd. 2006

Second printing, 2006

Publisher Cataloging-in-Publication Data (U.S.)

Haines, Tim.
The complete guide to prehistoric life/Tim Haines and Paul Chambers.
[216] p. : col. ill ; cm.
Includes index.
Summary: An illustrated guide to 112 beasts dating from the Cambrian Period to the Pleistocene Epoch, with profiles on physical characteristics, lifestyle, habitat, behavior and distribution across prehistoric earth.
ISBN-13: 978-1-55407-125-8
ISBN-10:1-55407-125-9
1. Vertebrates, Fossil. 2. Mammals, Fossil. 3. Dinosaurs. I. Chambers, Paul. II. Title.
566 22   QE841.H15 2005

Published in the United States by Firefly Books (U.S.) Inc.
P.O. Box 1338, Ellicott Station
Buffalo, New York 14205

Library and Archives Canada Cataloguing in Publication

Haines, Tim

The complete guide to prehistoric life/Tim Haines and Paul Chambers.
ISBN-13: 978-1-55407-125-8
ISBN-10: 1-55407-125-9
1. Animals, Fossil. I. Chambers, Paul, 1968- II. Title.
QE714.5.H33 2006 560 C2005-904257-5

Published in Canada by
Firefly Books Ltd.
66 Leek Crescent
Richmond Hill, Ontario L4B 1H1

Cover images
© BBC Worldwide Ltd.

Printed in Italy

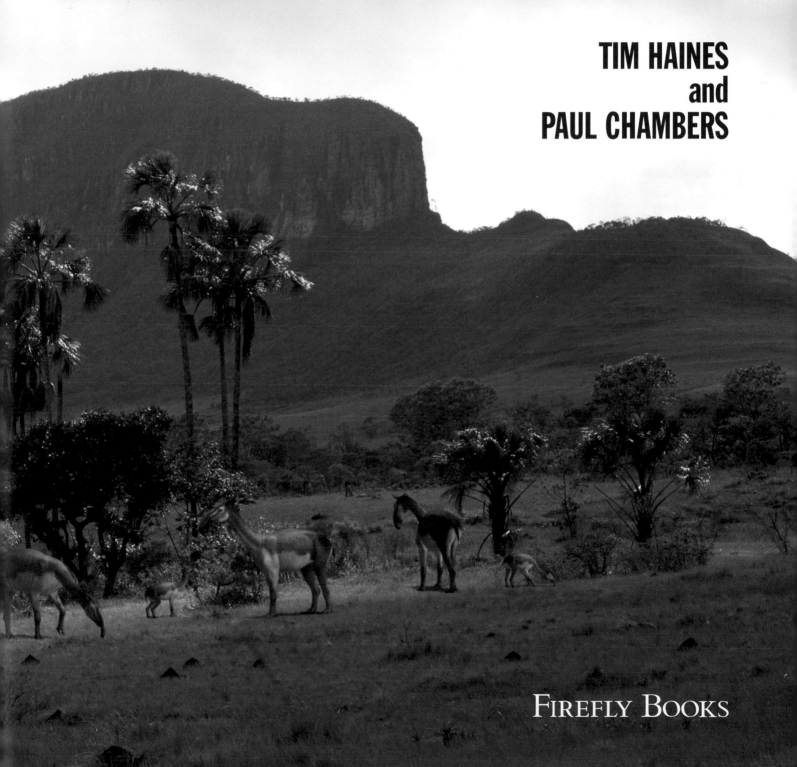

The COMPLETE Guide to
PREHISTORIC LIFE

TIM HAINES
and
PAUL CHAMBERS

FIREFLY BOOKS

Contents

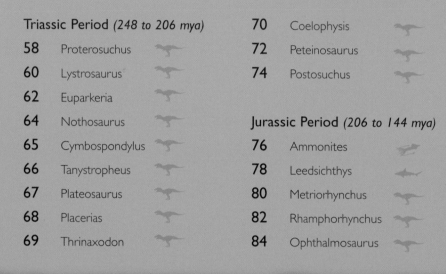

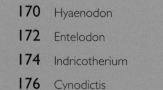
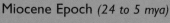

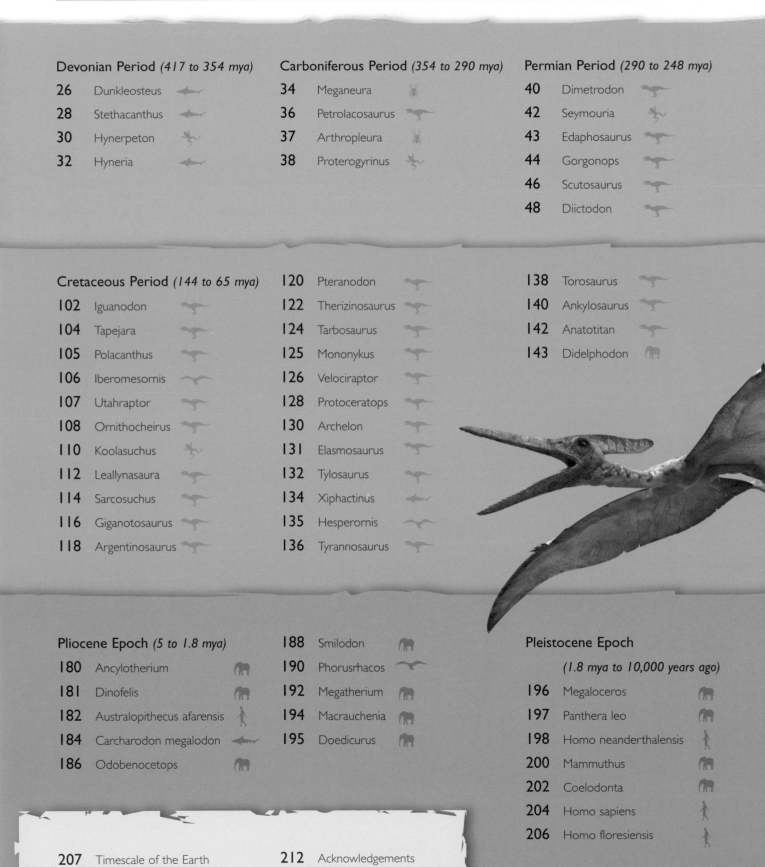

Introduction

Almost ten years ago we started making a television series called *Walking with Dinosaurs*. At the time it was very expensive and high risk because it aimed to use Hollywood technology to educate and inform rather than simply entertain. We offered the audience a vision of dinosaurs as real animals, not vindictive monsters. Fortunately the series broadcast to record audiences, and has since been seen by almost 400 million people worldwide.

However, we knew at the time that we were telling only half the story. Many weird and wonderful creatures came before and after the dinosaurs. The extraordinary success of *Walking with Dinosaurs* allowed us to complete this story. *Walking with Beasts* detailed the evolution of the mammals after the dinosaurs died out, and *Walking with Monsters* revealed the wonderful variety of creatures that thrived before the dinosaurs evolved. We finished with seven and a half hours of television that covered 4 billion years, cost millions of pounds and told the biggest story of them all – the evolution of life on Earth.

The combined coverage of these television series has allowed us to fulfil another long-standing ambition – namely, to produce a book that could tell the same story using the unique high-resolution stills made possible by the materials produced specially for the programmes. Across the following pages you will see careful re-creations of dozens of prehistoric animals, each sculpted by specialist modellers, scanned into a computer and then brought to life by talented animators.

Part of making these extinct animals look true to life was ensuring that we accurately portrayed the latest scientific thinking about their biology, behaviour and lifestyle.

Right > Together at last – using the latest digital techniques we have been able to range across time and bring back to life the weirdest collection of creatures that ever existed. Each has a different story to tell about the evolution of life on this planet.

At every stage over the last ten years we have depended on scientists to guide us about the look and feel of the animals we were re-creating. Much of this information is available in books and journals, but our desire for the latest and most up-to-date information often led us to talk to those scientists who routinely work with the fossils of specific animals. Over the years we have contacted more than 600 scientists, all experts in their field, and all of them major contributors to the sum total of our knowledge about the prehistoric world. Much of the information that appears in this book is thanks to them, and has never before been aired in public.

In addition to the computer-generated animals, the landscapes on to which they have been superimposed were also specifically chosen because their climate, plants and topography are a close match to the prehistoric environments in which our various animals lived. The background photographs in this book were taken in a wide range of exotic locations, from the jungles of Indonesia to the deserts of Utah and the coral reefs of the Red Sea and the Caribbean.

We believe that this book presents as accurate a vision of the world's prehistoric past as our technology presently allows, but the fossil record is far from complete, and new discoveries are constantly overturning established scientific 'truths': palaeontological thought does not stand still for long. All we can do is continue to refine our opinions and remain open to new evidence.

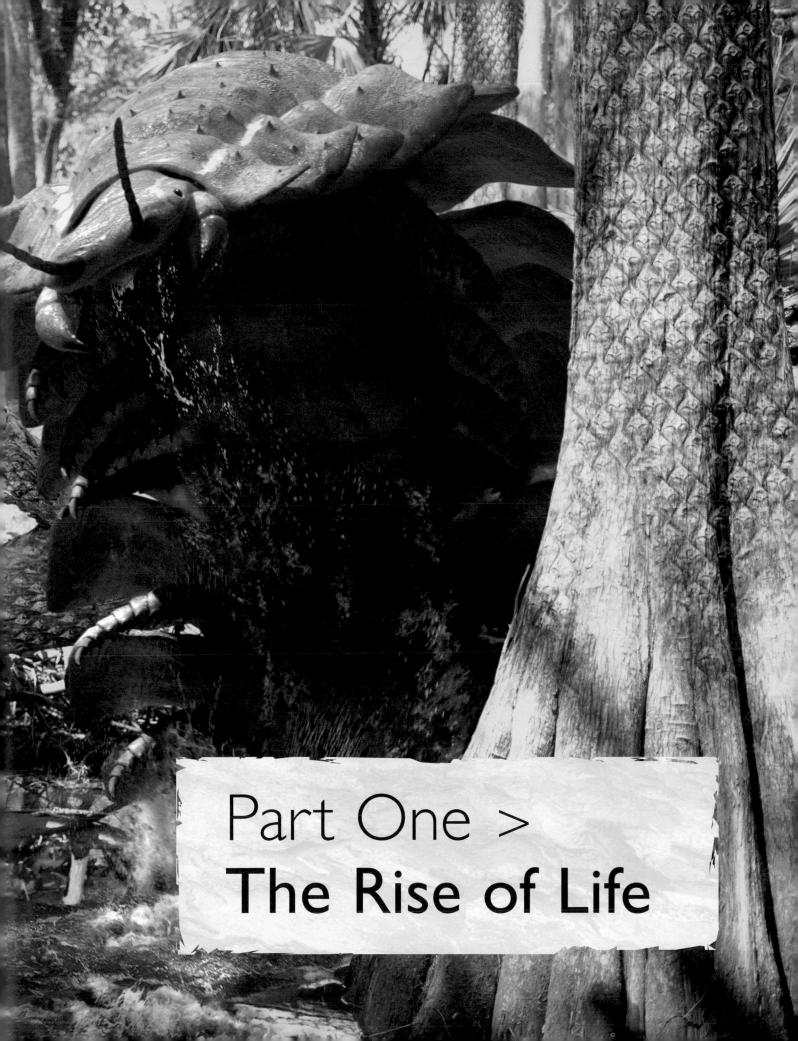

Part One >
The Rise of Life

The Precambrian

The Precambrian lasted for nearly 4 billion years and is the longest of the geological time periods. During this time the Earth went from being a lifeless ball to a home for many types of primitive life. However, the small and delicate nature of most life forms means that fossils from the Precambrian are very rare.

Archaean eon (4600–2500 million years ago)

For centuries it was believed that our planet Earth was at most only a few thousand years old and that all life had been created in just a few short days. Then, around 200 years ago, scientists began to make sense of the strange fossilized plants and animals that were being brought to them from quarries and rock outcrops around the world. Thanks to these fossils it is now known that the Earth is older than a few thousand years. In fact, it is a great deal older.

Studies of rocks, fossils, meteorites and other objects tell us that the Earth was formed around 4600 million years ago. Before this time there were no planets in our solar system; instead, there was just a mass of loose rock and dust swirling about our then primitive sun. Eventually the action of gravity drove the loose rock and dust together into large balls; these balls were the first planets, and they included the Earth.

For the first few hundred million years of its existence the Earth was a hellish place. Its surface was boiling hot, dominated by active volcanoes, and it was constantly pelted by giant asteroids. The atmosphere would have been thick and poisonous. Under conditions like these there was no way that any life form could have evolved.

One of the key events in the Earth's history occurred over 4 billion years ago, when a planet about the size of Mars slammed into it. The collision was so violent that part of the Earth was thrown into space, where it was moulded into the moon. The formation of our moon had a profound effect on the development of life because it stabilized the Earth's orbit around the sun, and created the tides, which helped aerate and cleanse the shallow seas where life first evolved. Oceans of liquid water spread around the globe. These offered a constant environment in which the slow process of life could begin.

Below > Over 4000 million years ago the Earth's surface was an unstable cauldron of fire, making it impossible for life to gain a foothold.

The oldest sign of life on Earth is a chemical signature in rocks from Greenland that are around 3850 million years old. No one knows for sure how life on Earth first evolved, and there are a number of different theories. One possibility is that the first simple organisms may have arisen in underwater hot springs, where there would have been an abundance of heat and the right sort of chemicals to permit life to evolve. For hundreds of millions of years life on Earth consisted of simple single-celled bacteria that formed thin layers of slime across the ancient seabeds.

The first organisms lived by extracting energy from chemical reactions, but then a group, the cyanobacteria, evolved that could extract energy from sunlight through photosynthesis. Bacteria formed vast mats on the ocean floor, sometimes creating layered mounds called stromatolites, the earliest fossils of which date from around 3500 million years ago. For hundreds of millions of years the slow pace of evolution meant that these simple organisms were the only life on Earth.

Proterozoic eon (2500–543 million years ago)

If the primitive Archaean eon marked the evolution of life on Earth, then the Proterozoic eon (meaning 'ahead of animal life') saw it develop from simple bacteria into complex plants and animals of the sort that we would recognize today.

At the start of the Proterozoic eon, around 2500 million years ago, the sea still harboured all life on Earth. The atmosphere contained significant quantities of carbon dioxide (and also possibly methane), but very little oxygen, which made the land simply too hostile for life. But this was soon to change.

The primitive bacteria in the Earth's seas produced oxygen as a by-product of their growth. This excess oxygen was released into the environment, and over millions of years it built up in the atmosphere until, around 2000 million years ago, the Earth had a permanently high level of this gas.

Oxygen is a highly reactive chemical and is an excellent fuel for life. Thus, the arrival of an oxygen-rich atmosphere sparked the evolution of new, complex single-celled organisms known as eukaryotes. Unlike bacteria (which are prokaryotes), each eukaryotic cell had a separate nucleus that contained DNA at its centre: it was a big step forward. The first eukaryote fossils, called acritarchs, are microscopic and date from around 1500 million years ago. Slowly these nucleated cells adapted to live together in colonies, eventually producing primitive sponge- and jellyfish-like creatures.

So towards the end of the Proterozoic eon the world was full of blind drifting jellies and static filter-feeders. Then, some believe, disaster occurred. Between 750 and 600 million years ago the Earth became engulfed in a severe ice age, which sent temperatures plummeting to -40°C (-40°F) and caused all the world's oceans to freeze over. This theory, which is still controversial, is sometimes called the 'Snowball Earth'. It is after the Snowball Earth, around 600 million years ago, that the first signs of complex animal life start to appear in the fossil record.

Fossils from 600 million years ago are very rare indeed, and they are often little more than faint impressions in the rock. Scientists find these fossils difficult to understand, but it is thought that they are the remains of simple animals. One of the most intriguing is *Spriggina*, a worm-like animal that must have pushed its way forward through the mud. Creatures like *Spriggina*, with a head and rear end, were to lay the foundation for an explosion in life forms to come, and represent the common ancestor to everything from the earthworm and crab to the dinosaur and whale.

These primitive animals all lived in warm, shallow coastal waters. Their fossils have been found around the world, and they are often called 'Ediacaran animals' because they were first found near Ediacara in Australia.

The Palaeozoic Era

Palaeozoic means 'ancient life', and it is at the start of this era that animals first began to have hard parts, such as shells and carapaces, in their bodies. Such hard parts fossilize well, and it is from this time onwards that scientists have been able to chart the rise and fall of individual groups of animals and plants.

Cambrian period (543–490 million years ago)

In the Cambrian period the fossils of many animals, such as the shelled **trilobites**, became common and they can be found at hundreds of locations across the world. Most significantly, this period saw the evolution of the complex eye – an organ that some palaeontologists believe helped accelerate the process of evolution because it led to the development of active hunters, which in turn drove prey to develop better defences.

During the Cambrian period the land was still a barren and hostile place, so all animal life lived in the shallow seas around the edge of the Earth's continents. Occasionally, huge underwater landslides would engulf these communities, burying them under tonnes of mud. These land-slides would preserve even the most delicate of soft-bodied animals as fossils, allowing us to see just what a strange place the Cambrian world really was.

From the rocks of the Canadian Burgess Shale (and other locations in China and Greenland), we know that bizarre animals, such as the giant predator *Anomalocaris*, swam through an alien landscape dominated by sponges and primitive seaweeds.

The Cambrian seas contained representatives from most of the major animal groups, including the arthropods (**trilobites**), molluscs (sea shells) and echinoderms (sea urchins, starfish). The oldest known vertebrate is also found in the Cambrian. It is a primitive fish, *Haikouichthys*, which lived 535 million years ago. This is our oldest known ancestor.

Above > A map of the world as it would have looked 535 million years ago.
Top right > Horseshoe crabs look little different to their Ordovician ancestors.
Right > Lichens may have been one of the first organisms to colonize the land.

Ordovician period

(490–443 million years ago)

Throughout the Ordovician period all animal life remained in the seas, which were home to primitive corals, sea urchins, starfish and sea shells, but the most populous creatures were the arthropods.

The **trilobites** were especially numerous, but they had been joined by the first chelicerates, the arthropod group that includes scorpions. One type of chelicerate, the sea scorpion *Megalograptus*, grew to giant proportions and was even capable of crawling on to land for brief periods of time. The land itself was barren, apart from a few species of primitive slime-mould and lichen that lived along stream banks.

In the seas was the orthocone *Cameraceras*, a relative of the modern squid, and at 10 m (33 ft) long the first truly gigantic swimming predator. Backboned animals were still represented by the primitive jawless fish that lived on the seabed, searching for small food fragments.

Silurian period (443–417 million years ago)

The Silurian world saw life progress at a steady pace. In the shallow tropical regions complex reef systems developed, built from corals, sponges and bryozoans. These reefs were home to smaller animals, such as jawless fish, sea lilies and brachiopod sea shells, but the arthropods still dominated life.

The sea scorpion *Pterygotus* reached a massive size, but there were also true scorpions, such as *Brontoscorpio*, which was capable of making short visits on to land. With so many large predators about, some jawless fish were now armour-plated and had evolved advanced senses.

It was near the end of the Silurian period that life first began to colonize the land in a meaningful way. The first plants, such as *Cooksonia* and some species of fungi, grew in clumps near to streams and rivers, but they reached a height of only 10 cm (4 in) or so.

Among these plants were the first land animals, which included primitive millipedes and other arthropods. Most of these animals were plant-eaters, but there were some predators as well.

Devonian period

(417–354 million years ago)

The Devonian period saw big changes both on land and in the seas. At the beginning of the Devonian life on land was still sparse and primitive, but only a few million years later the land was home to primitive forests, dominated by the first tree-like plant, named *Archaeopteris*, which grew in vast numbers alongside rivers and estuaries.

Animal communities on land were dominated by millipedes and predatory animals, such as the trigonotarbids, who were distant relatives of living spiders. It was during the Devonian that the first fish made the move from water to land, producing air-breathing, four-legged amphibians, such as **Hynerpeton**.

In the seas were two new types of swift and terrifying predator. The fish had come of age: with the evolution of a jaw, they developed the ability to tackle active prey, and quickly increased in variety and size. There were the sharks, such as **Stethacanthus**, whose sleek shape and sharp teeth made them formidable hunters. However, the biggest and most aggressive hunters were the giant placoderm fish, such as **Dunkleosteus**, which could reach lengths of 8 m (26 ft) or more. These were joined by the first bony fish, such as **Hyneria**, some of which were the ancestors of most of the modern fish in the oceans today.

Carboniferous period

(354–290 million years ago)

The Carboniferous period had a greenhouse climate that saw hot, humid conditions stretch from the Arctic to the Antarctic circles. Lowland areas had been colonized by thickly forested swamps dominated by tree-sized ferns and horsetails, and the gigantic, alien-looking lycopsid trees, some of which were 50 m (165 ft) tall.

Oxygen levels were very high, and may help explain why these flooded forests were home to an abundance of life that included giant arthropods, such as **Arthropleura**, and flying insects, such as mayflies and the giant dragonfly **Meganeura**.

The waterlogged conditions favoured the amphibians, such as **Proterogyrinus**, who could move and hunt in the streams and breed in the lakes. Although dominated by the amphibians, the Carboniferous also saw the evolution of the first reptiles, which were mostly small, lizard-like creatures, such as **Petrolacosaurus**. These little reptiles had eggs that could be laid away from water, something that helped lay the foundation for future success.

The Carboniferous seas were also teeming with life. The sharks and bony fish (the ancestors to most modern fish) dominated the oceans, while the seabed was home to complex coral reefs, some stretching for many kilometres along the ancient coasts.

Around 290 million years ago, at the end of the Carboniferous, the world was plunged into a global ice age. The temperature dropped, causing the tropical forests to shrink in size. In their place came vast ice sheets and glaciers, which spread outwards from the North and South poles, scouring the landscape. Many species could not cope with the change in climate, and in time became extinct.

Above > The Carboniferous swamp forests were home to thousands of species of exotic plants and animals. They were the ancient equivalent of today's rainforests.
Left > A map of the world as it would have looked 360 million years ago.

Permian period (290–248 million years ago)

The global ice age at the end of the Carboniferous period left the world a drier, cooler place. In the early Permian the tropical forests and swamps shrank and were replaced by open plains populated with scattered pockets of ferns and primitive conifers.

Amphibians, such as **Seymouria**, had previously dominated the Earth, but they needed to live close to water, so found the lack of tropical swamps hard going. In their place came the dry-adapted reptiles. They increased in number and size, producing animals such as **Dimetrodon** and **Edaphosaurus**, the Earth's first truly large land animals. The cold climate led to innovation among the reptiles, such as the large, heat-gathering sails on **Dimetrodon**.

By the late Permian period the world's continents had all joined together to form a gigantic landmass called Pangaea. In many parts of the world the climate was hot and dry with sparse rainfall, producing vast deserts. These deserts were home to the therapsids, an advanced breed of reptile that produced terrifying predators, such as **Gorgonops**, and burrowing plant-eaters, such as **Diictodon**. The therapsids dominated the landscape, but there were other large animals, such as the armoured **Scutosaurus**, a possible ancestor of the turtles, and the giant amphibian *Rhinesuchus*.

At the end of the Permian the climate started to become extreme, causing the deserts to expand and the coastlines to shrink. With fewer habitats available for plants and animals to live in, land and marine ecosystems came under sustained pressure, causing species to compete with each other for space and food: many became extinct. As conditions became worse, so the extinction of life accelerated until, by around 248 million years ago, up to 95 per cent of all species on Earth had died out. This was the largest extinction event in the known history of life on Earth, but from its ashes newer, more terrifying varieties of life would emerge.

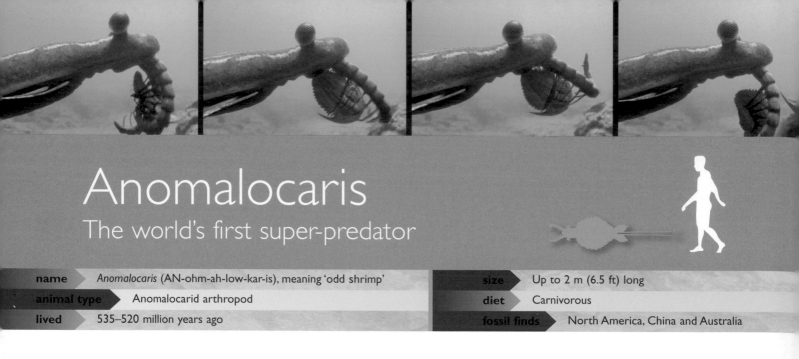

Anomalocaris
The world's first super-predator

name	Anomalocaris (AN-ohm-ah-low-kar-is), meaning 'odd shrimp'	size	Up to 2 m (6.5 ft) long
animal type	Anomalocarid arthropod	diet	Carnivorous
lived	535–520 million years ago	fossil finds	North America, China and Australia

At around the size of a man, *Anomalocaris* was the largest animal living in the ancient seas of the Cambrian period. It was a deadly predator that had a flexible, segmented body, large eyes and a circular mouth built from razor-sharp plates. It looked unlike any animal alive today; indeed, the Cambrian is famous for producing some very exotic animal designs that became evolutionary dead ends. *Anomalocaris* is one of these.

It probably lived in small shoals that would cruise the tropical coastal waters in search of prey. At the time around 90 per cent of all hard-shelled animals were arthropods (the group that today includes crabs, millipedes and insects); it is therefore not surprising that *Anomalocaris*, the top predator, was also an arthropod, although its shell was not as rigid as that of many of the smaller animals it was hunting.

Anomalocaris had large eyes mounted on stalks so that it could hunt other animals by sight. It could have eaten almost anything, but its favourite food was the **trilobite**, which crawled about on the seabed in great numbers. Most trilobites were protected with a hard shell. To overcome this *Anomalocaris* would use its curled front appendages to grab and hold the trilobite; it would then flex the trilobite backwards and forwards until the shell cracked, allowing it to get access to the soft meat inside.

Sometimes this giant predator would use the sharp interlocking plates of its circular mouth to take a bite out of a trilobite. In fact, several trilobite fossils have been found with a telltale wedge-shaped bite mark in their shells. As well as hunting, *Anomalocaris* would also have used its front appendages to rake the mud for soft-bodied animals, such as worms and small, water flea-like arthropods.

Anomalocaris swam by moving the flaps on the side of its body up and down in a wave-like motion. Compared to most Cambrian animals, it was manoeuvrable and capable of swimming at speed or just hovering in a single position. It has even been suggested that the evolution of *Anomalocaris* might have led to many other animals evolving hard shells in order to protect themselves from this voracious predator.

When scientists first found fossils of *Anomalocaris* they completely misunderstood its shape. The first

Top > *Anomalocaris* gets to grips with an unfortunate **trilobite**.
Left > Exquisitely preserved fossils, such as this one from China, have allowed scientists to study *Anomalocaris* in minute detail.

parts to be found were its long front appendages, discovered in 1892 and misidentified as belonging to a primitive type of shrimp. In 1911 fossils of the ring-shaped mouth were found, but these were misidentified as a type of jellyfish. Only in the 1980s was it realized that both the front appendages and mouth came from the same large animal.

Fascinating Fact > *Anomalocaris* was ten times longer than any other animal found from that time.

The first *Anomalocaris* fossils were found in the Burgess Shale of Canada, a rock formation that has produced many beautifully preserved Cambrian fossils, but near-complete fossils have since been found in several parts of the world, most notably in China and Australia. There are even fossils of its dung. Several species of *Anomalocaris* are known, but although most scientists think that it is a type of arthropod, its mouth and side flaps are not found in any other type of animal. Thus it is not known exactly how it is related to more familiar arthropods, such as **trilobites**, crabs and insects (e.g. *Meganeura*).

Trilobites
The Palaeozoic's big success story

name	*Trilobite* (TRY-lo-bite), meaning 'three lobes'		size	5 mm – 80 cm (0.2–32 in) long
animal type	Trilobite arthropod		diet	Scavenger
lived	520–248 million years ago		fossil finds	Worldwide

Trilobites may look like woodlice, but they are actually an extinct order of marine arthropods. They are among the most successful of all fossil groups, evolving around 520 million years ago and surviving through to the great extinction of 248 million years ago. There are over 15,000 described species, with the smallest being a few millimetres long and the largest, *Isotelus*, 80 cm (32 in) in length. Most trilobites lived on the seabed, searching for food scraps or small animals, but some were swimmers, while others burrowed their way through the muddy sediment.

Although they had a hard shell, trilobites were near the bottom of the food chain and were eaten by a whole range of predators. Many trilobite fossils have damaged shells where they were attacked by fish, orthocones and other arthropods, such as **Anomalocaris**. To protect themselves, some species could roll into a ball (like a woodlouse). Others lived in burrows, could swim, or had spines or knobbly shells. Trilobites were among the first animals to develop eyesight. Their eyes were complex and delicate, being

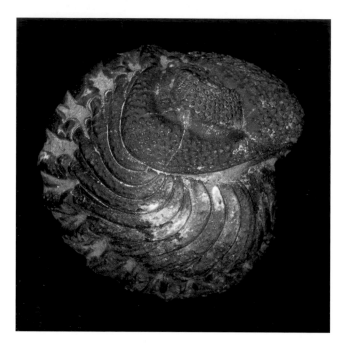

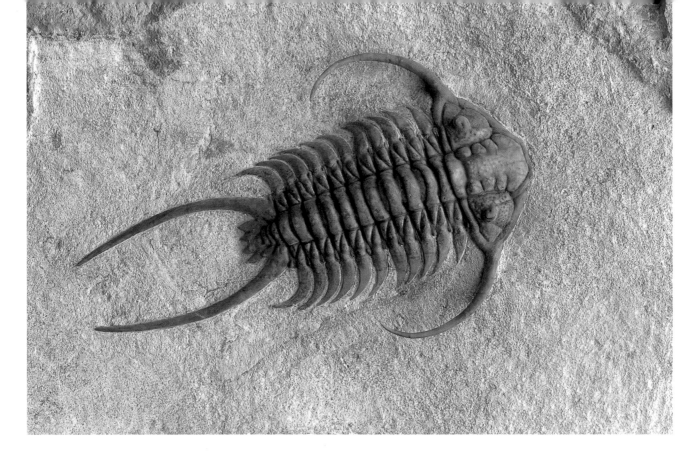

made from dozens of small crystal lenses, similar in design to those of modern insects. Trilobite eyes could not see as well as those of fish and other backboned animals, but they were sensitive to motion and would have helped the animal to avoid predators. Some species of burrowing trilobite did not need eyes, so they were blind.

Since trilobites lived on the Earth for over 270 million years, their fossils are extremely common. In some areas, such as North Africa, fossil trilobites are preserved in three dimensions, complete with their delicate spines and fragile eyes. Rock outcrops, such as the Burgess Shale in Canada, have preserved not only the trilobites' hard shell, but also the soft parts of their body, such as the legs and internal organs.

Fossil trilobites have been a source of fascination to humans for hundreds of years, and they were among the first extinct animals to be studied by scientists. Since we know of so many types of trilobite stretched out over millions of years, geologists can use individual species to work out the age of ancient rocks. If they find certain trilobite fossils they can date the rock to within a few million years.

For much of the Palaeozoic the trilobites were extremely common in the seas, but around 360 million years ago they became rarer, possibly because they were being eaten by newly evolved predatory fish, such as the sharks and placoderms (see *Stethacanthus* and *Dunkleosteus*). They eventually died out at the end of the Permian period, when a severe extinction event removed upwards of 95 per cent of all marine life. Despite their being among the commonest Palaeozoic fossils, trilobite specimens have a thriving market, with the best-preserved ones fetching huge sums.

Above > For much of the Palaeozoic trilobites would have been an extremely common sight, scuttling across the seabed or burrowing their way through mud.
Top > The delicate spines and other ornate features on trilobites have sometimes been preserved in incredible detail.
Top left > When threatened, some species of trilobite would curl up into a ball, a bit like a woodlouse.
Bottom left > Trilobites were among the first animals to evolve eyes which, although primitive, allowed them to see danger coming. It is possible that this development completely changed the course of evolution.

Fascinating Fact > Trilobites had eye lenses made of crystal.

Haikouichthys
The ancestor to all backboned animals

name		*Haikouichthys* (HI-koo-ICK-theez), meaning 'fish from Haikou'	size		2.5 cm (1 in) long
animal type		Agnathan (jawless) fish	diet		Scavenger
lived		535 million years ago	fossil finds		China

Haikouichthys was a jawless fish, whose primitive backbone allowed it to swim in a completely different way from the arthropods that lived around it. Its body was small, with no obvious fins, but it had eyes, gills, a nose, ears and a brain, making it one the most advanced animals in the ancient Cambrian seas. It was possibly hermaphroditic, meaning that it was both male and female at the same time. It laid eggs that would have hatched into larvae, and may have covered its skin in a layer of slime, just as its descendants, lampreys and hagfish, do.

Haikouichthys may have sought protection from predators such as ***Anomalocaris*** by living in shoals of several hundred fish, but its small size and body shape meant that it was not a great swimmer. It lived near to the seabed, and would eat by sucking in small particles of food through its jawless mouth.

The first *Haikouichthys* fossils were discovered in 1999 in Chengjiang, China. Since then, dozens of specimens have been found. The fossils are very well preserved, and show delicate features, such as the eyes and internal organs.

Haikouichthys is one of three species of jawless (or agnathan) fish to be found in the Early Cambrian period. These represent the oldest-known vertebrate (or backboned) animals on Earth, and they are therefore ancestral to all other backboned animals, including other types of fish, amphibians, reptiles and mammals. The exact origin of the vertebrates is not fully known, but the latest DNA research suggests that there was a split between vertebrate and invertebrate animals around 600 million years ago, in the Precambrian. Some species of jawless fish are still alive today, the most notable of which are the lampreys and hagfish.

Fascinating Fact > Fossils of *Haikouichthys* are so well preserved that you can still see their internal organs and slime glands.

20

Cameraceras

The largest carnivore in the Ordovician

name	*Cameraceras* (cam-er-a-sair-us), meaning 'chambered horn'	size	10–11 m (33–36 ft) long; shell 9–10 m (30–33 ft)
animal type	Cephalopod mollusc	diet	Carnivorous
lived	470–440 million years ago	fossil finds	North America

Cameraceras was the biggest animal alive during the Late Ordovician period. It was an orthocone – a squid-like animal that lived inside a long, straight shell. Orthocones are molluscs and distant relatives of the **ammonites** and the nautilus that exists today. Like them, orthocones had a shell that was divided into many empty compartments that it could fill with air or water, depending on whether it wanted to rise or fall in the water. The orthocone animal itself was quite small and lived in the last chamber of its shell. The largest orthocones, including *Cameraceras*, were longer than a bus and had tentacles almost 1 m (3.3 ft) long.

Swimming with such a large shell was problematic. To propel itself, the orthocone had a powerful fleshy tube (called a hyponome) that hung underneath its head. It would force water through this at great pressure, thus pushing itself in the opposite direction – the same principle as a jet engine. The hyponome was flexible and could be angled so that the orthocone could move in any direction, but it seems that most orthocones were best suited to travelling forwards.

Like most living cephalopod molluscs (a group that includes squid and octopuses), orthocones probably spent their daylight hours hiding in deep water.

At night they may have moved into shallower water to hunt fish, **trilobites** and sea scorpions (see *Megalograptus*). Their eyesight was poor, so they may have smelt out their prey using chemical sensors. With their eight tentacles (which had a grooved surface rather than the suckers of a modern squid) they would grab on to the prey and hold it, while a sharp beak (a bit like that of a parrot) slowly ripped the animal to pieces.

The word 'orthocone' refers to any straight-shelled cephalopod mollusc, and actually covers several different types of squid-like animal. Their fossilized remains have been found in their tens of thousands, and occasionally entire rocks are made from their remains. Some even have the colour of their shell preserved. Although a few species grew to gigantic proportions, most orthocones were only a few centimetres long. They were a major component of the primitive seas, first evolving around 495 million years ago and becoming extinct about 255 million years later.

Fascinating Fact > People used to think that fossil orthocone shells were unicorn horns.

Megalograptus
A giant spiny sea scorpion

name	*Megalograptus* (MEG-al-oh-grap-tuss), meaning 'giant writing'
animal type	Chelicerate (eurypterid) arthropod
lived	460–445 million years ago

size	1 m (3.3 ft) long
diet	Carnivorous
fossil finds	North America

Megalograptus was a sea scorpion, or eurypterid, belonging to a group of mostly aquatic arthropods that are related to horseshoe crabs. The sea scorpions were very hardy and could survive in almost any environment, including fresh water and even on land.

Like most sea scorpions, *Megalograptus* was well defended. It had a thick protective shell and a pair of long arms covered in sharp spines. Its only fear would have been attack from a larger sea scorpion or from a giant orthocone, such as ***Cameraceras***. Together these two groups of predators dominated the Ordovician and Silurian seas until the rise of the jawed fish in the Devonian.

When it needed to grow a sea scorpion would moult its shell, just as present-day crabs and scorpions do. The new shell underneath was soft and took several hours to harden. To minimize the risk of attack, hundreds of sea scorpions would come together in a mass to moult their shells. While their new shells remained soft, they would mate. The females were capable of storing the males' sperm for several months afterwards, allowing them to time their pregnancy to suit the environmental conditions.

Sea scorpions would spend most of their time under water, hunting for food. Some were ambush hunters, taking their prey by surprise. Others, including *Megalograptus*, would move along the sea floor, using the spines on their claws to feel for fish, **trilobites** or other animals living in the sand and mud.

Sea scorpions would swim, using their legs and tail to propel them, but they were also among the first animals to leave the sea and drag themselves on to land. Fossils of their drag marks have been found in many places. Since at this stage virtually nothing lived on the land, it is interesting to consider what could have enticed them out of the warm water. Perhaps they did so to mate or to feed on dead animals washed up on the beach.

Fossil sea scorpions have been found around the world. Most species were only 10–70 cm (4–28 in) in length, but some, such as ***Pterygotus***, grew to over 2 m (6.5 ft) long. The sea scorpions first evolved around 480 million years ago, but became extinct around 210 million years later, by which time some species had fully adapted to a life on land.

All sea scorpions are chelicerates, a group of arthropods that includes arachnids (spiders, mites and scorpions, e.g. ***Brontoscorpio***) and the xiphosures (horseshoe crabs). The chelicerates are an ancient group of arthropods, with their oldest fossils dating to about 500 million years ago in the Cambrian period. The evolutionary relationship between the main types of chelicerate is not fully understood, and the question of whether the scorpions and sea scorpions are descended from one another is especially controversial.

Cephalaspis
A heavily armoured fish

name	Cephalaspis (kef-al-ASS-piz), meaning 'head shield'	size	45–60 cm (1.5–2 ft) long
animal type	Agnathan (jawless) fish	diet	Detritivorous
lived	425–385 million years ago	fossil finds	Worldwide

Cephalaspis was a sturdy, muscular fish that was about the same size as the average trout. It shared many features with modern fish, including a backbone, scaly body, eyes and fins, and was much more advanced than the primitive *Haikouichthys* from which it was descended. Even so, *Cephalaspis* had many primitive characteristics, such as its heavy armour and small tail.

Covering the front part of its body, *Cephalaspis* had hard plates that protected it from attack but made it very heavy and thus unable to swim very fast or far from the seabed. On its head and running down the side of its body *Cephalaspis* had special sensory organs. These were capable of generating an electrical field that could detect the presence of nearby objects and animals. Using these organs, *Cephalaspis* could navigate its way at night and through murky water. It could also have detected the presence of any predators, and kept out of their way.

Having no jaws meant that *Cephalaspis* could not bite or chew its food. Instead, it would live by grubbing in the mud for detritus, such as loose particles of plants and meat, plus any worms, crustaceans and

other small animals. There is evidence that some species of *Cephalaspis* migrated from the shallow sea into freshwater rivers and lakes, perhaps to mate and lay eggs as salmon and eels do. It is thought that *Cephalaspis*, or a very similar fish, may have been the ancestor to jawed fish (such as *Hyneria*), and thus also the ancestor to all land vertebrates, including humans.

Around 400 million years ago the jawless (agnathan) fish began to decline sharply, probably because they were outcompeted by new types of faster fish, such as the sharks and placoderms (see *Stethacanthus* and *Dunkleosteus*). By the end of the Devonian period (354 million years ago) the jawless fish were very rare, but they did not become extinct altogether, and can still be found today in the form of the slimy, eel-like hagfish and lampreys.

Fascinating Fact > *Cephalaspis* could generate an electrical field to detect predators.

Brontoscorpio
The biggest scorpion of all time

name	Brontoscorpio (BRON-tow-SKOR-pee-oh), 'thunder scorpion'	size	94 cm (3.1 ft) long
animal type	Arachnid (scorpion) arthropod	diet	Carnivorous
lived	420–415 million years ago	fossil finds	Europe

Scorpions have changed little since they first evolved around 430 million years ago. The oldest-known species lived in the sea or in rivers, and some grew to be very large indeed. *Brontoscorpio* was the largest of all – the size of a medium-sized dog.

Brontoscorpio looked little different from most present-day scorpions. It had eight legs, large claws, a hard shell and a poisonous stinger. However, its large body had to be supported by the water, so much of its life was spent in the sea. To breathe under water it had gills, something that modern scorpions lack. It had good eyesight and could have located food by sight. It would use its claws and stinger to catch worms, fish, **trilobites** or any other animals that came within range. Food would have been broken into small pieces by the claws and fed into the mouth bit by bit.

Although rather slow and cumbersome, *Brontoscorpio* was capable of making short journeys on to land. It would have come ashore to avoid predators after moulting its shell. It would wait on land for several hours while the new shell hardened, after which it could make its way back to the sea again.

The oldest-known fossilized scorpions are around 430 million years old, but it was not until 340 million years ago that scorpions started to live on land permanently. *Brontoscorpio* is known from only a few fossilized fragments found in 1972 in England; smaller but more complete fossilized scorpions are known from rocks of the same age. *Brontoscorpio* is an arachnid and thus related to the spiders and mites. The arachnids are a subgroup within the chelicerates, a group of arthropods that includes the sea scorpions (e.g. **Megalograptus** and **Pterygotus**) and horseshoe crabs. The evolutionary relationship between the scorpions and the chelicerates is not known, but some people think that they may be closely related to the sea scorpions.

> **Fascinating Fact** > *Brontoscorpio* had a stinger about the size of a small light-bulb.

Right > A 1-m (3.3-ft) long *Brontoscorpio* is attacked by the gigantic sea scorpion **Pterygotus**. It was perhaps to escape giant marine predators such as these that the first animals (such as millipedes and scorpions) began to move from the seas and on to the land.

Pterygotus
The largest arthropod ever

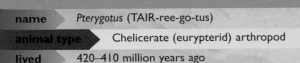

name	Pterygotus (TAIR-ree-go-tus)	size	2.8 m (9.2 ft) long
animal type	Chelicerate (eurypterid) arthropod	diet	Carnivorous
lived	420–410 million years ago	fossil finds	Worldwide, except Antarctica

Pterygotus was the largest of the sea scorpions (or eurypterids) ever to evolve (see **Megalograptus**), and was one of the top predators in the sea. It lived in shallow coastal areas, hunting fish, **trilobites** and other animals by stealth. It would have ambushed its prey by partially burying itself in the sand. Then, when a fish or other unwitting animal came within range, *Pterygotus* would rise up and grab it with its claws.

Unlike some other types of sea scorpion, adult *Pterygotus* were far too large to be able to crawl on land, so they lived permanently in the sea. Their fossils, however, have been found in a wide range of habitats, including freshwater lakes that would have been many kilometres from the sea. It has been suggested that they could have migrated into these lakes by swimming upstream from the sea, possibly in pursuit of fish.

Pterygotus was an accomplished swimmer and could move with speed and agility through the water. It would swim by flapping its long, flat tail up and down; the broad, flat part at the end would push it through the water in much the same way as the fluke on a whale's tail does. It would steer and stabilize itself using its legs.

Fossils of *Pterygotus* are relatively common, although complete skeletons are rare. It was one of the last of the gigantic sea scorpions: later species were much smaller and more nimble. The decline of the larger sea scorpions may be related to their relative slowness and vulnerability to attack. They would, for example, have been an easy catch for a large predatory fish, such as **Dunkleosteus**.

Around 330 million years ago some species of sea scorpion were so pressurized by predators that they left the water altogether and started to live on land. One species, named *Megarachne*, lived in the flooded swamp forests of the Carboniferous period, where it would eat small invertebrates or food detritus. For many years the two known fossils of *Megarachne*, both from Argentina, were mistaken for those of a giant spider with a legspan of 50 cm (20 in). Indeed, *Megarachne* was officially the largest known spider of all time until, in 2004, the fossils were correctly identified as land-dwelling sea scorpions. After the Carboniferous the demise of the swamp forests meant that the sea scorpions as a whole became very rare until, around 270 million years ago, they became extinct altogether. Their nearest living relations, the horseshoe crabs, continue to thrive in today's oceans.

Dunkleosteus
A Devonian shark killer

name	*Dunkleosteus* (dunk-lee-owe-stee-us), meaning 'Dunkle's bones'
animal type	Placoderm (armoured) fish
lived	370–360 million years ago

size	8–10 m (26–33 ft) long
diet	Carnivorous
fossil finds	North America, Europe and North Africa

Named after Dr David Dunkle, an American palaeon-tologist who studied its fossils, *Dunkleosteus* could grow to almost 10 m (33 ft) long, making it the largest pred-ator in the Devonian sea. It was an alien-looking ani-mal, with a head that was encased in solid and inflexi-ble armour-plating, and a muscular body that was streamlined – much like that of a shark.

The secret of *Dunkleosteus*'s success was its mouth. The evolution of a movable lower jaw, around 420 million years ago, marked the start of the fishes' rapid rise to dominance in the seas. *Dunkleosteus* was a mag-nificent example of this. Instead of teeth it had two razor-sharp shearing plates made from bone, which would flash past each other in a scissor-like motion. The jaws were powerful, and anything caught between them would have been sliced clean in two. This allowed it to hunt other large animals, killing them and then eating them with an efficiency that jawless animals could not hope to match. The evolution of the jawed fish coincides with the decline of many other marine animals, including the **trilobites**, sea scorpions (see **Pterygotus**) and jawless fish (see **Cephalaspis**). It is likely that the jawed fish outcompeted all these animals.

Dunkleosteus would have pursued its prey with a sudden burst of speed, before crippling or killing them

Above > A beautifully preserved *Dunkleosteus* fossil.
Top > *Dunkleosteus* moves in for the kill with speed and agility.

with its shearing plates. Some fish defended themselves with large spines on their dorsal fins, which would stick in the roof of *Dunkleosteus*'s mouth, preventing the fish from being swallowed. At least one *Dunkleosteus* died after getting such a fish stuck in its throat.

The lack of teeth meant that it couldn't chew and so had to swallow large pieces of food in their entirety. Sometimes the stomach couldn't cope with such large chunks of meat, and *Dunkleosteus* would be forced to vomit on to the seabed. Balls of this fossilized vomit are common in Late Devonian rocks.

Fascinating Fact > Fossilized sick balls produced by *Dunkleosteus* have been found.

Dunkleosteus was a placoderm fish, which means that some parts of its body, especially the head, had a covering of armour-plating. Most placoderms were quite small and lived by scavenging. However, *Dunkleosteus* was part of an elite group of placoderms known as the arthrodires. There are over 200 known species of arthrodire, most of which were swift predators. The placoderms first evolved around 420 million years ago from jawless fish (e.g. **Haikouichthys**) and quickly dominated the Devonian seas. They were highly versatile and could live in estuarine and freshwater environments as well as in the sea. However, the placoderms' success was short-lived, and the group as a whole became extinct around 355 million years ago, possibly because they were outcompeted by the more nimble sharks.

Stethacanthus
One of the first sharks

name	Stethacanthus (STETH-ack-anth-uss), meaning 'chest spine'	size	0.7–2 m (2.4–6.5 ft) long
animal type	Symmoriiform shark	diet	Carnivorous
lived	370–345 million years ago	fossil finds	North America, Europe and northern Asia

The first sharks evolved around 410 million years ago and quickly became successful predators, so by the Late Devonian period (around 360 million years ago) they were a common sight in coastal waters. Their success has continued through to the modern day. Apart from the strange-shaped fin on its back, *Stethacanthus* was little different from present-day sharks. With their streamlined bodies, wide mouths, sharp teeth and smooth swimming action, these early sharks had obviously stumbled upon an unbeatable design.

Stethacanthus was a predator, but its relatively small size would have restricted its diet to small fish, cephalopods and perhaps some types of arthropod, such as **trilobites**. It would have been capable of swimming at moderately fast speeds, and would have hunted by chasing small fish or grabbing animals from the seabed, as many modern reef sharks do.

The most obvious feature on *Stethacanthus* is its strange-shaped dorsal fin, which has a wide, flattened top covered in hundreds of rough, tooth-shaped scales. A patch of the same rough scales was also found on the top of its head.

Only males had this strange-shaped fin, which means that it probably played a role in mating, possibly as part of a courtship display or to intimidate rivals. However, it has also been suggested that when viewed from certain angles the fin and the rough-shaped scales could make *Stethacanthus* look as

Fascinating Fact > *Stethacanthus* is also known as the 'ironing board' shark because of its weird dorsal fin.

though it had a gigantic, tooth-laden mouth. This might have served to scare off potential predators. Nonetheless, even though *Stethacanthus* was a reasonable-sized shark for its time, it would have stood little chance if confronted by some of the giant placoderms (e.g. **Dunkleosteus**) with which it shared the seas.

Stethacanthus was a symmoriiform shark, an ancient shark lineage that first evolved around 370 million years ago, but did not become extinct until around 300 million years ago. In some parts of the world *Stethacanthus* fossils – which are usually teeth because the rest of the body, being made of cartilage, did not preserve well – are found in great numbers. It is possible that these locations may have been seasonal breeding grounds, such as sheltered bays, to which hundreds of *Stethacanthus* migrated in order to mate and give birth.

A few *Stethacanthus* fossils have been preserved in extraordinary detail, showing the outline of the animal and some of its biological detail. This has allowed scientists to deduce exactly what it would have looked like when alive, and even to see the sex of the animals. Despite the demise of *Stethacanthus*, sharks are an evolutionary success story and have changed little during their 410-million-year history.

Below > Well-preserved fossils of *Stethacanthus* have allowed scientists to see the sex of individual animals. From this it has been deduced that only the males had the 'ironing board' fin.

Hynerpeton
A primitive landlubber

name	Hynerpeton (HI-ner-pet-on), meaning 'creeping animal from Hyner'	size	2 m (6.5 ft) long
animal type	Primitive amphibian	diet	Carnivorous
lived	360 million years ago	fossil finds	North America

Hynerpeton is a primitive amphibian (the group that contains frogs and salamanders) that lived in the rivers and estuaries of the Late Devonian period. It was one of the first vertebrate (backboned) animals able to leave the water and venture on to land. It is from animals like *Hynerpeton* that all other backboned land animals are descended: this includes reptiles such as the dinosaurs, and mammals such as ourselves.

The body of *Hynerpeton* was well adapted to living in water, being streamlined, with a long, powerful tail for swimming. However, it also had four legs and feet (each with eight toes), air-breathing lungs, and eyes that could see in air. These allowed it to leave the water and move around on shore, something that no other large animal could do. In addition, *Hynerpeton* would have needed specialized internal organs, such as kidneys, to help it cope with life on land. These would have been used to help it get rid of waste products from the blood, such as nitrogen and carbon dioxide.

Despite these features, *Hynerpeton* was not comfortable out of water. On shore it was slow and lumbering, with its legs barely able to support its weight and its lungs only just able to provide it with enough oxygen to survive (to overcome this its skin was also used to absorb oxygen). But the land did provide *Hynerpeton* with many advantages. It was underpopulated and thus free from large predators. It also had an abundant and easy-to-catch food supply in the form of arthropods, such as scorpions and millipedes. The land was also a safer place to lay eggs than the water, as there were fewer animals to steal them. Like modern amphibians, *Hynerpeton* would probably have laid a mass of jelly-like spawn in a shallow pool or damp hollow, and then left it to develop into tadpoles.

Even so, *Hynerpeton* would have spent most of its time in the water, where it could swim with ease and use its legs and feet to pull itself along the bottom or to clamber through vegetation. It was a carnivore, and probably an ambush predator, hiding itself in weed or among rocks so that it could lurch out at passing fish and arthropods. It might have been able to detect nearby prey using special electrical organs that ran down the side of its body. Periodically, it would have to return to the surface, where it would gulp down lungfuls of air before diving again.

Above > Out of water *Hynerpeton* would have been slow and awkward, but with no large predators about it was safer to be on land than in the water.

Left > Fossilized footprints such as these are proof that around 360 million years ago large amphibians were beginning to make journeys away from the water's edge.

Incomplete fossils of *Hynerpeton* have been found in the Catskill Formation of Pennsylvania, which, 355 million years ago, was a tropical estuary system, with deep river channels and wide muddy banks – ideal for an animal that could live in and out of the water. *Hynerpeton* is just one of several primitive amphibians from the Late Devonian whose fossils, including footprints, have been found in places as far apart as Greenland and Australia. These early amphibians are sometimes collectively called primitive 'tetrapods', which simply means 'four legs'.

The study of primitive amphibians such as *Hynerpeton* has long fascinated scientists, as the move from water to land is deemed to be a key moment in the evolution of life. Although the topic is still a controversial one, it is thought that these early amphibians are descended from lobe-finned fish, such as **Hyneria**, whose stout fins evolved into legs and their swim bladder into lungs. It is still not known whether *Hynerpeton* is the direct ancestor to all later backboned land animals (including humans), but the fact that it had eight fingers, not five, suggests it is simply our evolutionary cousin.

Fascinating Fact > *Hynerpeton* had eight digits on each foot.

31

Hyneria
A 2-tonne killer fish

name	*Hyneria* (Hi-ner-ree-ah), meaning 'from Hyner'
animal type	Rhipidistian (lobe-fin) fish
lived	360 million years ago

size	4 m (13 ft) long
diet	Carnivorous
fossil finds	North America

Hyneria was a lobe-finned fish – it had a solid, bony skeleton, stout, muscular fins and was covered in large scales. It also had powerful jaws and sharp teeth, and could swim fast, making it a deadly predator. It used keen eyesight and an acute sense of smell to detect prey, such as large fish and even primitive amphibians, such as **Hynerpeton**. Victims would be caught in its jaws and then, while still alive, swallowed head first.

The hunting grounds of *Hyneria* were the estuaries and rivers of the Late Devonian period. As well as cruising in deep water, *Hyneria* could venture into very shallow water, something that was useful in tidal estuaries or in rivers that would seasonally become shallow. Once in shallow water, *Hyneria* would use its powerful front fins to haul itself about, searching for larger fish stranded by the tide or amphibians struggling to get out of the water.

It has even been suggested that *Hyneria* could have beached itself for short periods, using its air bladder (an organ that helped it float in the water) as a primitive set of lungs. Gulping air may also have allowed *Hyneria* to hunt in murky or stagnant water, where a lack of oxygen might hinder normal fish. Indeed, scientists believe they have found evidence that some lobe-finned fish were remarkably active on land, despite their size. Fossils have been found which suggest that some species of these fish may have deliberately stranded themselves on the top of steep mud-banks. Here they would wait, then slide down into the water below, grabbing at passing prey with their powerful jaws.

The first *Hyneria* fossils were found near to the town of Hyner, Pennsylvania, in 1968. Many specimens have been found since then, although a complete skeleton has yet to be discovered. *Hyneria* was just one of many species of lobe-finned fish that were common in the Late Devonian period, the most famous species being *Eusthenopteron*, whose well-preserved fossils are common and so have been intensively studied by scientists for decades. These fish varied in size, from a few centimetres to several metres in length.

Right > Giant predatory fish, such as *Hyneria*, would have regularly eaten sharks, which were mostly small during the Devonian period.

Fascinating Fact > Despite its unwieldy size, *Hyneria* could drag itself on to land to search for food.

The lobe-finned fish first evolved around 415 million years ago, and for a long period of time they were successful large predators. However, during the Carboniferous period (354–290 million years ago) they declined in importance, possibly because the ray-finned fish (or actinopterygians, a group that includes most modern fish, such as goldfish and trout) started to become common.

Today the lobe-fins are represented by their close relative the coelacanth, a large, deep-sea fish that was only discovered in 1938 off the coast of South Africa.

Lobe-finned fish, such as *Hyneria* and *Eusthenopteron*, have been heavily studied because it is thought that they are the ancestors to primitive land animals, such as **Hynerpeton**.

Meganeura
An eagle-sized dragonfly

name	*Meganeura* (MEG-an-you-rah), meaning 'giant nerves'
animal type	Odonatid (dragonfly) insect
lived	311–282 million years ago

size	75-cm (2.5-ft) wingspan
diet	Carnivorous
fossil finds	Europe

The tropical, swampy forests of the Late Carboniferous period were home to many species of flying insect, the largest and most deadly of which were the dragonflies. Of these, *Meganeura* was the biggest species; it may even have been the largest flying insect in geological history. It has been suggested that its gigantic size came about as a response to increased levels of oxygen in the Carboniferous atmosphere. This would have made the air denser, and thus easier to fly in, allowing the muscles to work more efficiently.

An accomplished flier, *Meganeura* spent most of its time in the air, flapping or gliding its way through the forest in search of food. It would hunt on the wing, plucking other insects, including dragonflies and mayflies, from the air. It could also take small, ground-dwelling animals by swooping downwards and grabbing them with its legs. It rarely stopped flying, and may have landed only in order to mate and lay its eggs. The larvae, which would have been around 30 cm (12 in) long, were also voracious predators. They would have lived in vertical burrows on water-logged land, and would have emerged periodically to hunt for spiders, insects and small amphibians. Once caught, the prey would have been dragged back to the burrow and eaten.

The biology of the Carboniferous dragonflies is little different from that of present-day species. The only significant difference is the location of the reproductive organs: on Carboniferous dragonflies these were right at the end of their tail, whereas living dragonflies have them nearer the head. The fact that present-day dragonflies have the potential to eat one another during mating has led to great speculation about how ancient dragonflies, with their different reproductive organs, might have mated. One idea is that the male (who is most at risk of being eaten) could have put the female into a trance-like state and then flipped her on to her back before mating; he could then have escaped to a safe distance before she revived again.

Meganeura was one of many species of giant dragonfly in the Carboniferous and Early Permian periods, but not all of these were carnivorous: fossils of their mouth parts show that some ate only fruit. Around

Fascinating Fact > Even *Meganeura* larvae – 30 cm (12 in) long – were ambush predators.

290 million years ago the world became a drier and cooler place, causing the forests to shrink in area and the dragonflies to become much smaller. Even so, they were able to adapt to changes in the Earth's climate and actually became more numerous.

Insects, including the dragonflies, are one of the greatest success stories in the history of life. The oldest fossilized insects are 400 million years old. Back then they were only a few millimetres long, although a recent discovery suggests that even at this stage some species could have had wings and flown. By 330 million years ago insects could fly, and by the Late Carboniferous the forests teemed with crawling, burrowing and flying insects. Some of these insects, such as the mayflies, dragonflies and cockroaches, are still with us. Today insects make up an estimated 95 per cent of all life on land.

Petrolacosaurus
One of the oldest reptiles

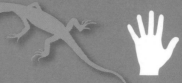

name	Petrolacosaurus (PET-ro-lak-co-saw-rus)	size	40 cm (16 in) long
animal type	Diapsid reptile	diet	Carnivorous
lived	300–290 million years ago	fossil finds	North America

The Carboniferous saw the evolution of the first reptiles, which included *Petrolacosaurus*. Although primitive when compared to later reptilian giants, such as the dinosaurs, *Petrolacosaurus* and other similar species were much more versatile than amphibians, such as **Proterogyrinus**.

The small and lizard-like *Petrolacosaurus* lived in the dense Carboniferous swamp forests, where it would scamper along the ground searching for insects and grubs. In these dense swamps reptiles were rare and frequently eaten by large predatory amphibians: there was no suggestion that one day they would evolve into giants that would dominate the planet.

Petrolacosaurus had many features that made it more fully adapted to living on land than amphibians. For a start, it did not need to keep its skin wet, and because it laid eggs with hard shells it did not need to return to the water in order to breed. These were great advantages, as they meant that *Petrolacosaurus* did not even need to live near water.

The eggs of *Petrolacosaurus* were probably large (compared to its body size), and the shell would have protected the embryo within. However, the reptile embryo absorbed air through the shell, making the eggs vulnerable during times of flood: if covered with water, the embryo would drown. For this reason *Petrolacosaurus* probably made its nest on high ground, which would also have had the advantage of hiding it from amphibian predators.

The oldest reptile fossils come from Scotland and are around 340 million years old. For millions of years reptiles remained rare and it was only after the world became cooler and drier (around 290 million years ago) that the reptiles' ability to survive away from water gave them an advantage over the semi-aquatic amphibians. By the Late Permian period (256–248 million years ago), they were the largest land animals on Earth.

Petrolacosaurus fossils were first found during the 1930s in Kansas, but they were poorly preserved and it was difficult for scientists to see how the creature could be related to other fossil reptiles. Better specimens found in the 1970s showed that this small reptile is actually a probable ancestor to the archosaurs, the group that includes the dinosaurs and crocodiles.

Arthropleura
The largest land arthropod of all time

name	Arthropleura (AH-throw-ploo-ra), meaning 'side joint'	size	0.3–2.6 m (1–8.5 ft) long
animal type	Athropleurid arthropod	diet	Omnivorous
lived	340–280 million years ago	fossil finds	North America, Middle East and Europe

Arthropleura looked like a giant centipede, and had a body made of up to 30 individual plates, underneath each of which was a pair of legs. The surprising thing about *Arthropleura* is its size. It is the largest-known land arthropod, with the biggest species growing far longer than a man is tall.

What *Arthropleura* ate is a matter of debate among scientists, as none of the fossils have the mouth preserved. However, it is reasonably certain that it would have had a sharp and powerful set of jaws. Based on this assumption, it used to be thought that *Arthropleura* was carnivorous, but recently discovered fossils have been found with pollen in the gut, suggesting that the creatures ate plants. It is probable that smaller *Arthropleura* species were vegetarian, while the largest ones were omnivorous, using their jaws to tackle vegetation, as well as to hunt small animals and insects. It is estimated that the average *Arthropleura* could have eaten its way through a tonne of vegetation a year.

Fossilized footprints from *Arthropleura* have been found in many places. These appear as long, parallel rows of small prints, which show that it moved quickly across the forest floor, swerving to avoid obstacles, such as trees and rocks. When moving at speed, its body would stretch and become longer, giving it a greater stride length and thus allowing it to move faster.

As it moved about, *Arthropleura* would have brushed against many different types of plant, and may have helped the forest reproduce by moving pollen or spores about the place. It is also thought that *Arthropleura* was capable of travelling under water, and that it may have returned to lakes and rivers in order to moult its shell. This would have made it vulnerable to attack by large fish and amphibians. On land an adult *Arthropleura* would have had few enemies.

There are around ten species of *Arthropleura*; their fossils are common and widespread, and have been found in rocks across Europe and North America. Complete fossils are rare, with most specimens consisting only of a few body segments.

The oldest arthropleurids, the group to which *Arthropleura* belongs, date from 420 million years ago, and they may have been among the first arthropods to live permanently on land. The exact relationship between the arthropleurids and the myriapods (centipedes and millipedes) is not known. The oldest

species are small, and it was not until around 325 million years ago that the first gigantic forms appeared. The arthropleurids became extinct around 280 million years ago, probably because the lowland swampy forests in which they lived became very scarce around this time.

Proterogyrinus
A crocodile-like amphibian

name	*Proterogyrinus* (PRO-tuh-roe-gee-ree-nus), meaning 'early wanderer'	size	2.3 m (7.5 ft) long
animal type	Anthracosaur amphibian	diet	Carnivorous
lived	325–320 million years ago	fossil finds	North America

During the Late Carboniferous period (323–290 million years ago) the amphibians were an exciting and diverse group of animals that had come to dominate the waterlogged forest world. *Proterogyrinus* was one of the largest amphibians, and was perfectly adapted for life in the swamps. It was a top predator that hunted both on land and in water. Its powerful jaws had sharp teeth and could handle quite large animals, such as fish, reptiles and other amphibians.

Most Carboniferous amphibians were good swimmers and could move at speed through the rivers and lakes surrounding the lowland forests, but few were capable of walking on land. Most either stayed in the water or could only wriggle and crawl their way across the muddy riverbanks. *Proterogyrinus* was different: it was a good swimmer, but it could also move on land with relative ease, using its stout legs to walk rather than crawl. Being able to hunt away from the water's edge meant that

Left > The sorts of tree found in Carboniferous forests were very different from those we see today. Some types, such as the lycopsids, whose bark is pictured here, grew tall and straight like telegraph poles and could reach heights of 50 m (160 ft) in only a few years. Much of the coal we burn on our fires is made from the fossilized remains of trees like these.

Proterogyrinus could catch food in places where its rivals couldn't go. It also meant that it could escape water-borne enemies, such as predatory fish, by clambering on shore, or over the logs and other obstacles that frequently choked the swamps' river channels.

Fossilized tadpoles from a close relative of *Proterogyrinus* have been found, and suggest that it probably mated, laid eggs and developed into adults in much the same way as present-day amphibians. *Proterogyrinus* would have needed to lay its eggs in water, but it is possible that being able to move on land could have allowed it to use damp environments, such as hollow logs or isolated pools, instead of ponds and rivers where the eggs were vulnerable to predation.

Proterogyrinus was an anthracosaur, which means that it belongs to a group of advanced amphibians with reptilian characteristics; consequently they are sometimes referred to as reptilomorphs. It is likely that the first reptiles, such as **Petrolacosaurus**, evolved from reptilomorph amphibians. The amphibians with which we're familiar today (frogs, newts and salamanders) evolved from a separate group (the batrachomorphs), whose ancestors also lived in Carboniferous swamps. After the Carboniferous large amphibians became rarer until, by the start of the Mezozoic era, 248 million years ago, their fossils were virtually unknown. The last truly gigantic amphibian in the fossil record is **Koolasuchus**.

Left > Being a large predator on land was an advantage in the Carboniferous forests as there were few other animals to compete with.

Right > Insects were so big during the Carboniferous it is possible that amphibians, such as *Proterogyrinus*, were occasionally insectivores.

Dimetrodon
The famous sail-backed carnivorous reptile

name	Dimetrodon (DIE-met-roh-don), meaning 'two-measure teeth'	size	0.5–3.5 m (1.6–11.5 ft) long
animal type	Synapsid (pelycosaur) reptile	diet	Carnivorous
lived	282–256 million years ago	fossil finds	North America, Europe and Russia

Dimetrodon was a big, mobile predator whose evolution heralded the beginning of the age of large reptiles. Unlike amphibians, reptiles had waterproof skin that could hold the moisture within their bodies – a big advantage in the arid climate of the Permian period.

The most obvious feature of *Dimetrodon* was the large sail on its back. This was not just for display, but also to help control its body temperature. *Dimetrodon* was cold-blooded, which meant that its body heat came from the sun. The sail increased its body surface area by nearly 50 per cent, thus providing more skin for the sun to heat. On cool mornings *Dimetrodon* would have been cold and sluggish, and so it would turn its sail towards the rising sun. The heat boost this provided would speed up *Dimetrodon*'s metabolism, allowing it to hunt around an hour ahead of most other animals. In the heat of midday, *Dimetrodon* could turn its sail away from the sun, thus helping it to cool off and avoid overheating.

Dimetrodon probably hunted using its acute senses of smell and sight. It would have pursued its prey on foot, although it may have been able to hunt fish by standing in shallow streams and rivers. It usually lived in

Below > Commonly mistaken for a dinosaur, *Dimetrodon* lived millions of years before them and is more closely related to mammals.

lowland, swampy regions, but could survive in a range of environments, including harsh mountain areas.

Dimetrodon was one of the first animals to have teeth of different sizes, which allowed it to catch and eat different types of animal. Strong jaw muscles operated the jaws, and as well as allowing it to kill quickly, they could also chew its food, making digestion quicker and more efficient.

Dimetrodon fossils were first discovered in the United States in 1878, but they have since been found in Europe and Russia. They show that *Dimetrodon* species varied in length from a few centimetres to 3 m (10 ft).

Dimetrodon was a pelycosaur, a group of reptiles with primitive mammalian characteristics, such as teeth of different sizes. They are forerunners of more advanced therapsid (mammal-like) reptiles, such as **Gorgonops**.

Seymouria
A land-living amphibian

name	Seymouria (SEE-moor-ee-ah), meaning 'animal from Seymour'		size	90 cm (3 ft) long
animal type	Labyrinthodont amphibian		diet	Carnivorous
lived	282–260 million years ago		fossil finds	North America and Europe

The dry climate of the Permian suited the reptiles better than the amphibians, but *Seymouria* was one amphibian that almost beat the reptiles at their own game. It had many reptilian characteristics, including long and muscular legs, dry skin and an ability to conserve water. It could also excrete excess salt from its blood through a gland in its nose. All of this meant that *Seymouria*, unlike other amphibians, could live away from water for extended periods, an ability that allowed it to move about the landscape in search of insects, small amphibians and other prey.

Male *Seymouria* had thick skulls and probably had rowdy mating contests, battering one another with their heads until a victor emerged. After mating, the females would have returned to the rivers and ponds to lay their eggs. The larvae would then develop in the water, hunting for worms and insects until they were big and strong enough to survive on land. *Seymouria* lived in semi-arid regions, suggesting that they could cope on land better than most amphibians.

Fossils of *Seymouria* were first found in Seymour County, Texas, and presented an immediate puzzle. Scientists noticed that the body was very reptilian, but that the skull appeared amphibian. For decades no one was sure to which group *Seymouria* belonged. It was eventually deduced that its biology was more amphibian than reptilian.

Seymouria is part of the seymouriamorpha subgroup of amphibians, but it is often described as being a 'reptilomorph'. This means that, like **Proterogyrinus**, it is part of the group of amphibians from which the reptiles evolved. The first species of seymouriamorpha evolved around 290 million years ago, at the start of the Permian period, but by the Late Permian they were all extinct.

Fascinating Fact > Among the excellent fossil records of *Seymouria* are some of its tadpole larvae.

Edaphosaurus
A spectacular Permian herbivore

name		*Edaphosaurus* (ED-aff-oh-saw-rus), meaning 'pavement lizard'
animal type		Synapsid (pelycosaur) reptile
lived		282–256 million years ago

size		3 m (10 ft) long
diet		Herbivorous
fossil finds		North America and Europe

At first glance, *Edaphosaurus* looks very similar to **Dimetrodon**, but the two animals are only distantly related, and were very different from one another. The most obvious difference was that while **Dimetrodon** was a voracious meat-eater, *Edaphosaurus* ate only plants.

Until the this time, all land reptiles and amphibians had been carnivores or insectivores. This was because meat is high in protein and fat, easier to digest and easy to find. Plants, on the other hand, are tough to eat and nutritionally poor, which means that a huge amount of them must be consumed in order to extract enough energy to live on.

By the Permian, however, some reptiles, such as *Edaphosaurus*, had evolved a range of features to help them tackle the tough land plants. *Edaphosaurus* had a battery of peg-like teeth that could tear at leaves and branches, and jaw muscles that would help to grind its food. Yet more teeth were embedded in the roof of its

Fascinating Fact > *Edaphosaurus* was the first-known plant-eating reptile to evolve.

mouth for mashing the plants before they reached the stomach. To help it digest the plants *Edaphosaurus* had a big body; this allowed it to eat a large volume of food in one go, and gave it a long gut that helped it extract the maximum nutrients from its poor diet.

The sail on *Edaphosaurus* was used to keep its body temperature constant, and, like **Dimetrodon**, it was a pelycosaur reptile. Although it preferred to live in or near water, *Edaphosaurus* could also survive in quite harsh climates, such as arid deserts and cold mountains. Large *Edaphosaurus* would have had few enemies, although old and injured ones would have fallen prey to predators, such as **Dimetrodon**.

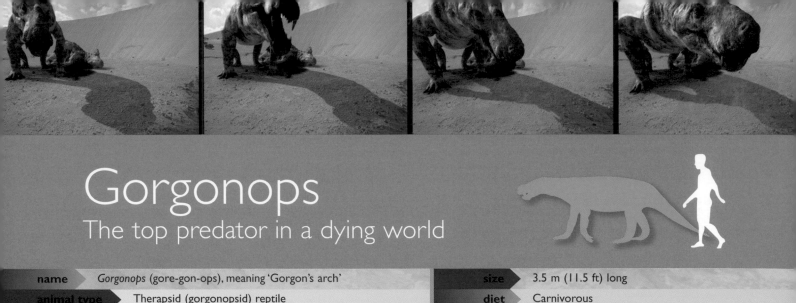

Gorgonops
The top predator in a dying world

name	*Gorgonops* (gore-gon-ops), meaning 'Gorgon's arch'	size	3.5 m (11.5 ft) long
animal type	Therapsid (gorgonopsid) reptile	diet	Carnivorous
lived	252–248 million years ago	fossil finds	Africa

Top > *Gorgonops* had acute eyesight and a great sense of smell. This helped it to seek out and stalk up on its prey.
Above > This mummified *Gorgonops* shows its fearsome 12-cm (5-in) long fangs.
Below > Having two sizes of teeth was advantageous: the fangs could be used to attack and kill prey while the small teeth could rip and tear meat away from the victim's body.

Gorgonops was a gigantic carnivorous reptile about the size of a rhinoceros, but with a sleek, wolf-like body and long, sharp teeth that permitted it to chase and bring down animals that were much larger than itself. It lived during the Late Permian period (256–248 million years ago), which was a time of crisis for life on Earth.

Increasingly high temperatures and low rainfall saw the creation of deserts across the world, driving many species towards extinction. It was in these harsh conditions that one particular group of reptiles called therapsids flourished. They quickly evolved into many different species, but although they grew large they were nothing to do with the dinosaurs: they were, in fact, more closely related to mammals.

Like most therapsids, *Gorgonops* was perfectly adapted for life in a hot, semi-desert environment. It had long, upright legs that allowed it to walk and run efficiently, and its jaws were powerful and full of many sharp teeth. In addition, it had two 12-cm (5-in) sabre teeth that could be used to kill prey quickly. These, combined with its size, good eyesight and sense of smell, made *Gorgonops* the top predator of its time.

Gorgonops probably lived a life similar to that of many present-day large predators, hunting in the mornings and afternoons, and resting in the shade during the heat of the day. Being the top predator, it was capable of taking on most other animals. It would use its excellent eyesight and sense of smell to seek out potential prey, which, when spotted, would be chased and brought down with its teeth. Whether *Gorgonops* lived in social groups is uncertain, but it is thought that it would have taken several of them working together to bring down a large reptile, such as a pareiasaur (e.g. **Scutosaurus**).

There were many varieties of therapsid, some of which were carnivorous, others herbivorous (see **Diictodon** and **Lystrosaurus**). The therapsids were important because although they were reptiles (and thus cold-blooded), they had many mammalian

characteristics, such as long legs that were tucked underneath the body, and specially shaped teeth that could deal with different types of food. It was from the therapsids that the first true mammals evolved (see **Thrinaxodon**).

Gorgonops is known only from fossils found in South Africa, but fossils of other gorgonopsids, the group to which it belonged, have been found around the world. As a group, the gorgonopsids were large carnivores, most of which looked broadly similar to *Gorgonops*. There were many species of gorgonopsid, some of which were only the size of a small dog, but others, including *Gorgonops*, were rhinoceros-sized.

Fascinating Fact > *Gorgonops* was the first of many predators through the ages to sport long, curved canines, known as sabre teeth.

The gorgonopsids first evolved around 260 million years ago and quickly spread themselves around the globe. Despite their large size, acute senses and success as predators, they did not survive the great extinction event that engulfed the world 248 million years ago, when most of the animals they hunted became extinct.

Scutosaurus
A giant desert herbivore

name	*Scutosaurus* (scoot-o-saw-rus), meaning 'shield reptile'		size	3.5 m (11.5 ft) long
animal type	Anapsid reptile		diet	Herbivorous
lived	252–248 million years ago		fossil finds	Europe (Russia)

The giant reptile *Scutosaurus* was one of the strangest-looking animals from the Permian period. Its bulky body and face were covered in bony lumps and protrusions, which to our eyes make it look aggressive and ugly. It had short, elephant-like legs, and a solid, muscular body. Most of these features would have helped protect it from attack, and also made it look threatening, even to equally large predators, such as ***Gorgonops***.

As a plant-eater living in a semi-arid climate, *Scutosaurus* would have wandered widely in order to find fresh foliage to eat. It may have stuck closely to the riverbanks and floodplains where plant life would have been more abundant, straying further afield only during times of drought. Its teeth were flattened and could grind away at the leaves and young branches before digesting them at length in its large gut. Given that it needed to eat constantly, *Scutosaurus* probably lived alone, or in very small herds, so as to avoid denuding large areas of their edible plants.

With its large cheekbones, *Scutosaurus* may have been able to make a loud bellowing sound. It had excellent hearing and could have heard another animal bellowing from some distance away. These noises could have been used for mating or as warning signals.

Despite its relatively small size, *Scutosaurus* was heavy, and its short legs meant that it could not move at speed for long periods of time, which made it vulnerable to attack by large predators. To defend itself *Scutosaurus* had a thick skeleton covered with powerful muscles, especially in the neck region. Underneath the skin were rows of hard, bony plates (scutes) that acted like a form of chain mail. These would have made *Scutosaurus* a difficult target to hunt, but two or three

Above > Fossils of *Scutosaurus* reveal that it was one of the best-defended animals living in the Permian period.

Gorgonops hunting together could conceivably have exhausted and brought down an adult.

Scutosaurus belonged to a group of reptiles known as the pareiasaurs, most of which were large and heavily built animals, around 2–3 m (6.5–10 ft) in length. Their skulls show some similarities to those of the first fossil turtles, and it is speculated that the two may have been directly related to one another. It has been

Fascinating Fact > Despite its appearance, *Scutosaurus* may be an ancestor of turtles and tortoises.

suggested that through time the bony plates underneath *Scutosaurus*'s skin could have become larger, then fused together to form the hard shell of the turtles.

Diictodon
The gopher of the Permian

name	Diictodon (DY-ik-toe-don), meaning 'two teeth'	size	45 cm (1.5 ft) long
animal type	Therapsid (dicynodont) reptile	diet	Herbivorous
lived	256–252 million years ago	fossil finds	Africa and Asia

The small, cat-sized *Diictodon* was a very common sight on the arid plains of the Late Permian period. It had a large head, a beak (but no teeth), a barrel-shaped body and a short tail. It lived in deep, narrow, corkscrew-shaped burrows that led down to a final chamber about 1.5 m (5 ft) underground. These burrows may have been the key to *Diictodon*'s success, as their special design would both keep the interior cool and prevent predators from being able to dig them out.

Each burrow housed only a single pair of *Diictodon* (a male and a female), which has raised the possibility that these animals might have mated for life, something that is relatively rare in the animal kingdom. *Diictodon* were abundant, and in the same small patch of ground there could be literally dozens of burrows, although (unlike modern rabbit warrens) none of them were interconnected. The final chamber was lined with vegetation and used by the female to produce and rear her young.

Diictodon had powerful legs and sharp claws, as well as a sharp, horny beak. The burrows were dug using only the front claws (not the horny beak), with the loose earth being pushed out behind. There is evidence of differences between the males and females in *Diictodon*, the males possessing a large pair of tusks and

Below > These male and female *Diictodon* were curled up with one another when their nest was flooded and they died.

Fascinating Fact > *Diictodon* are missing links: reptiles on the way to becoming mammals.

being larger in size. This could mean that there was rivalry between males, with occasional confrontations over territory or mating rights. Like some present-day burrowing animals, the male might also have made the nest as part of the mating ritual.

Diictodon had excellent senses of balance, sight and smell, and was nimble, darting in and out of its burrow to gather food. As a plant-eater, it used its beak to crop off leaves, roots and branches; it might also have been able to dig for buried roots and tubers.

During times of crisis, such as drought, living underground would have helped *Diictodon* survive by providing a cool, damp environment. It could also have allowed it to feed on any underground roots that protruded from the tunnel walls. However, *Diictodon*'s habit of building its burrows on floodplains or in river-banks sometimes had tragic consequences, when the waters periodically rose and flooded the nests, drowning the animals within and filling the tunnels full of sediment. Fossils of the unfortunate victims are common, and often reveal that the male and female *Diictodon* were cuddled up against one another when the flood waters struck.

To say that *Diictodon* was a common animal is an understatement; its fossils account for around half of all backboned animals found in South Africa from the Late Permian period. This has made it one of the most studied and best-understood animals from this period of time. Many fossilized examples of its burrows have also been found.

Diictodon was widespread, both in time and geographically. It was a therapsid reptile (see **Gorgonops** for an explanation of this), but within that group belonged to a subgroup known as the dicynodonts. All the dicynodonts were vegetarian and varied in size from a few centimetres to 2 m (6.5 ft) or more in length. The first dicynodonts evolved around 260 million years ago and were abundant during the Late Permian. They were among the few reptile groups to survive the great extinction event that took place 248 million years ago, and went on to dominate the earliest part of the Triassic period (see **Lystrosaurus**) before being outcompeted by other larger reptiles.

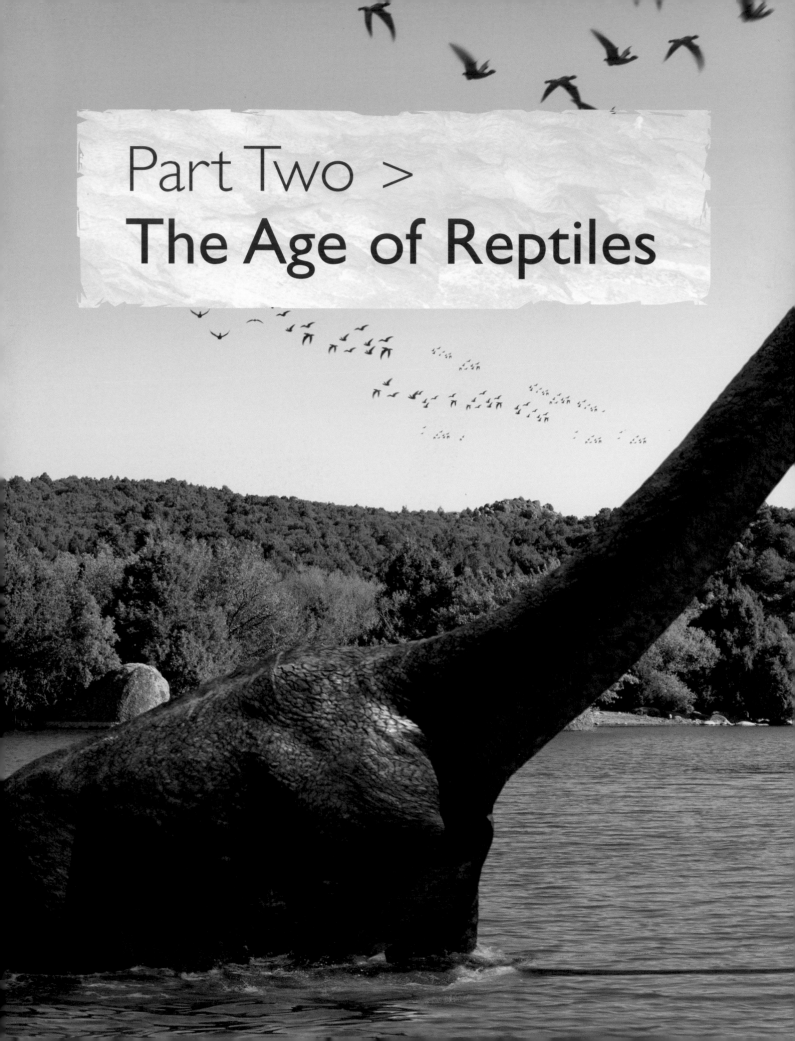

Part Two >
The Age of Reptiles

The Mesozoic Era

The Mesozoic (or 'middle life') era is sometimes called the 'age of reptiles' because during this time the dinosaurs and their relatives evolved into the dominant animal group on Earth. However, there was much more to the Mesozoic than its reptilian killers. It was a time of great innovation and diversity among life on Earth.

Triassic period (248–206 million years ago)

At the beginning of the Triassic period the world was still recovering from the terrible mass extinction event that affected the Earth at the end of the Permian. On land the climate across much of the globe was hot and dry, but there was enough seasonal rainfall to sustain many species of plant. Especially common were primitive conifers, ferns and gingkos, the fossilized remains of which have been found around the world, including the polar regions.

During the Early and Middle Triassic (248–227 million years ago), the animals that survived the Permian mass extinction found themselves in a world with few competitors and almost no large predators. Herbivorous animals, such as *Lystrosaurus*, took advantage of this and could be found in large numbers, as could predators such as *Proterosuchus*. Most animals underwent a phase of rapid evolution, producing new and interesting species, such as *Euparkeria*, a possible forerunner to the dinosaurs. The Early Triassic saw the reptiles re-enter the water, producing the semi-aquatic *Nothosaurus*.

Certain events in the Late Triassic period (227–206 million years ago) were to set the scene for much of the rest of the Mesozoic era. The gigantic supercontinent of Pangaea began to break apart, creating several smaller continents. Up until the Late Triassic the land had been dominated by the mammal-like (therapsid) reptiles, such as *Placerias*

Above > 245 million years ago the world was dominated by the supercontinent Pangaea.
Right > Two herbivorous *Placerias* stand and face the giant predator *Postosuchus* in the dry open forestland of the Late Triassic world.

Above > Araucaria conifers were very common across many Triassic and Jurassic landscapes; nowadays they are restricted to only a handful of southern hemisphere locations.
Top right > During the Jurassic period the giant continent Pangaea began to break apart, creating new seaways, oceans and coastlines that quickly began to teem with life.
Bottom right > The Late Jurassic saw the evolution of many types of large reptile, including *Diplodocus*, a giant species of sauropod dinosaur that specialized in grazing the vast fern prairies.

and *Lystrosaurus*, in addition to several groups of disparate but unusual reptiles, such as *Tanystropheus* and *Proterosuchus*. However, within a short space of time the mammal-like reptiles declined in number (apart from cynodonts, such as *Thrinaxodon*, which went on to give rise to the true mammals). In their place came the archosaur reptiles, which consisted of three important groups that would go on to dominate the Earth. These were: the dinosaurs, such as *Coelophysis* and *Plateosaurus*; the pterosaurs, such as *Peteinosaurus*; and the crocodilians, such as *Postosuchus*. In the seas the reptiles were also making their mark, producing the gigantic ichthyosaur *Cymbospondylus*.

The end of the Triassic period is marked by another mass extinction event that affected life both on land and in the oceans, but to nowhere near the same degree as the one at the end of Permian period. The cause of this extinction event is unknown. It was once thought that the asteroid that produced the huge Manicouagan crater in Canada (100 km/62 miles in diameter) might be responsible, but it turned out to be too old.

Jurassic period (206–144 million years ago)

The Early Jurassic period (206–180 million years ago) saw the global climate become warm and humid. In the polar regions conifer forests became widespread, while the tropics saw the spread of conifers, ferns and cycads. As the continents slowly continued to pull apart, so monsoon conditions occurred across some lowland regions, creating wide river basins that were prone to flash floods. It was during this time that the dinosaurs and pterosaurs established themselves, producing more species, becoming larger and more numerous, and spreading themselves about the globe. The same was also true of the marine reptiles, such as the ichthyosaurs and plesiosaurs, as well as shellfish, such as the **ammonites**.

The Middle and Late Jurassic period (180–144 million years ago) saw some tropical parts of the world become drier and more arid. Perhaps in response to this change in climate, many dinosaur species started to become truly gigantic in size. The herbivorous sauropods produced heavyweight monsters, such as **Diplodocus** and **Brachiosaurus**, while the carnivorous theropod dinosaurs produced multi-tonne killers, such as **Allosaurus**. Around these lived other dinosaurs, such as **Stegosaurus** and **Othnielia**. The flying pterosaurs produced fish-eating species, such as **Rhamphorhynchus**, and tiny insectivores such as **Anurognathus**.

The warm Jurassic seas became rich in plankton, which in turn fed large fish, such as **Leedsichthys**. The predatory plesiosaurs were represented by the long-necked **Cryptoclidus** and the gigantic **Liopleurodon**, while the first marine crocodilians, such as **Metriorhynchus**, hunted in the shallow seas.

Cretaceous period (144–65 million years ago)

The Cretaceous period was characterized by a largely warm and equable climate, with seasonal rainfall that allowed abundant plant life to exist from the Equator to the polar regions. The flowering plants (angiosperms) with which we are so familiar today first evolved in the late Jurassic, but as the Cretaceous progressed, so they became more dominant across the world. By the end of the Cretaceous the flowering plants had replaced the conifers, ferns and cycads in many environments, laying the foundations for their domination in the later Cenozoic world.

The continents also continued to pull away from one another, creating new seas and oceans (most notably the Atlantic), and at times preventing land animals from moving freely about the globe. Gradually the different continents began to evolve their own unique plants and animals.

The Cretaceous was a time of giants. South America saw the evolution of **Giganotosaurus** and **Argentinosaurus**, the largest known land animals, while North America was home to sturdy giants, such as **Tyrannosaurus** and **Torosaurus**. The dinosaurs also became more specialized, with dinosaur species such as **Velociraptor** and **Protoceratops** surviving in the Mongolian desert sand-dunes, and **Leallynasaura** close to the South Pole. In the background were the mammals (**Didelphodon**, for example), which remained small, but whose numbers began to increase markedly, especially towards the end of the Cretaceous.

Above > Compared to the dry and drought-ridden Jurassic, the Cretaceous was a much warmer and wetter world. The evolution of new types of plants, especially the flowering plants (or angiosperms), led to a greener, more verdant environment. These conditions provided new food sources for browsing dinosaurs but they also gave cover to their predators. Here an allosaurid dinosaur has killed a species of **Iguanodon** that had been feeding by a riverbank.

In the seas there was also a change, as previously dominant predators, such as the ichthyosaurs and pliosaurs, gave way to swift predatory fish, such as *Xiphactinus*, and a new breed of giant reptile – the mosasaurs – which included *Tylosaurus*.

In the air the pterosaurs grew large, with species such as *Ornithocheirus* and *Pteranodon* being able to glide long distances, possibly even across continents. The pterosaurs were joined by small, primitive birds, such as *Iberomesornis*, although there were also varieties of large, flightless seabirds, such as *Hesperornis*.

The end of the Cretaceous, 65 million years ago, saw a mass extinction event that removed an estimated 40 per cent of the known animal families on Earth. (This extinction event is sometimes called the 'K/T event' because it marks the boundary between the (K)retaceous period and Tertiary era.) Its most famous victims were the dinosaurs (although their descendants, the birds, survived), but many fossil groups also died out, including the pterosaurs, ammonites and mosasaurs. Many other animal groups were very badly affected, but managed to survive.

What caused the end-Cretaceous extinction is still a matter of debate. A strong contender is the impact of a giant asteroid, for which there is much evidence. However, in recent years studies have revealed that the pattern of extinction is far more complicated than was at first thought, with many animal groups (including the dinosaurs) declining in the 2 or 3 million years prior to the asteroid impact. This decline coincides with a marked cooling of the world's climate. Many scientists now believe that the K/T event might have been caused by a combination of prolonged environmental stress, which weakened many animals, followed by the catastrophic asteroid impact and the several months of global darkness and freezing weather that followed it. Whatever the real cause, 65 million years ago saw the sudden termination of the Mesozoic era and the 'age of reptiles' and the start of the Cenozoic era and the 'age of beasts'.

Proterosuchus
A long-legged crocodile ancestor

name	*Proterosuchus* (PRO-ter-oh-sook-us), meaning 'early crocodile'		size	3.5 m (11.5 feet) long
animal type	Basal archosaur reptile		diet	Carnivorous
lived	248–245 million years ago		fossil finds	Africa and Asia

Proterosuchus was the largest land animal on Earth during the Early Triassic period, but in modern terms it was only a little bigger than a Komodo dragon. With long jaws, powerful neck muscles, short legs and a lengthy tail, it looked a bit like a primitive crocodile, and in fact did have a lifestyle that is broadly similar to the Nile crocodile of modern-day Africa.

The long, muscular tail on *Proterosuchus* was excellent for swimming, and could push it through the water at speed. However, the animal also had stout legs that enabled it to walk comfortably on land. Being able to move between the land and the water was a great advantage, and enabled *Proterosuchus* to control its body temperature by sunbathing or cooling off in the water. It also increased the number of animals that it could hunt.

Like some present-day crocodiles, *Proterosuchus* was probably a 'sit and wait' predator. This meant that rather than going in search of prey, it would have stayed in the same place all year round, waiting for

Fascinating Fact > *Proterosuchus*'s hook-shaped mouth made it almost impossible for any animal to escape from its jaws.

animals to come to it. Doing this would have saved a great deal of energy and lessened the need to eat so often. In fact, *Proterosuchus* might have been able to survive for several months without food.

Although it could live and swim in the water, *Proterosuchus* preferred to hunt land animals rather than fish. Its eyes were located on the top of its head, allowing it to hide just under the surface of the water, where it would wait for animals to come and drink. When close enough, *Proterosuchus* would spring upwards and drag its victim into the water, drowning and then eating it.

Fossils of *Proterosuchus* were first found in 1903 in South Africa, but have since been found in Russia and China. It was an early example of an archosaur, a large group of reptiles that included the crocodiles, dinosaurs, pterosaurs and birds. Although the archosaurs were around in the Late Permian period (256–248 million years ago), it was only after the extinction of giant predators such as **Gorgonops** that they began to grow bigger and more widespread. Fossils of early archosaurs are rare, which makes *Proterosuchus* very valuable in scientific terms. Palaeontologists think that it may be an ancestor to the crocodilians, a group that has managed to survive through to the modern day (see **Postosuchus**, **Metriorhynchus** and **Sarcosuchus**).

Top > *Proterosuchus* almost catches a small *Euparkeria*.
Right > Two *Proterosuchus* have mounted an underwater attack on a *Lystrosaurus* that was trying to swim across a river.
Left > Having eyes on top of its head meant that *Proterosuchus* could lurk just under the water without being seen. The hooked mouth gave it an extra-strong grip on its prey.

Lystrosaurus
A highly successful pig-like reptile

name	*Lystrosaurus* (LISS-trow-saw-rus), meaning 'shovel reptile'	size	2 m (6.5 ft) long
animal type	Therapsid (dicynodont) reptile	diet	Herbivorous
lived	248–245 million years ago	fossil finds	Antarctica, Africa, Europe and Asia

It took the world many millions of years to recover from the great extinction event that occurred at the end of the Permian period. The landscape lost the majority of its large land animals, but those that did survive found themselves in a world with few rivals for food and even fewer large predators. Of the survivors, the therapsid (mammal-like) reptiles were among the first to bounce back, and of these *Lystrosaurus* was by far the most successful.

For a period of a couple of million years, *Lystrosaurus* was the most abundant large animal on Earth. It could be found almost everywhere from the Antarctic to the Arctic, and in some areas dominated the landscape, moving in large herds that would have stripped the leaves and seeds from all the low-lying trees and plants. Its tusks may have been used to dig roots and tubers from the ground.

Lystrosaurus was one of the largest animals on Earth at that time, but even so, it was still only the size of a pig. It had excellent eyesight and a good sense of smell and hearing, useful for detecting predators such as **Proterosuchus**, whose fossilized remains are sometimes associated with those of *Lystrosaurus*.

The first *Lystrosaurus* fossils were discovered in South Africa in the mid-nineteenth century, but they have since been found all over the world (except in the Americas). It was initially thought that *Lystrosaurus* must have been aquatic, living in rivers and lakes, a bit like a hippo. But recent fossil discoveries make it clear that while it liked to live near water, it probably didn't spend much time in it. Some people also suggested that it might have created deep burrows, but this is also thought unlikely.

The discovery of *Lystrosaurus* fossils in so many different parts of the world gave scientists strong evidence that the world's continents were once all joined together. When, in 1969, the first *Lystrosaurus* fossils were found in the Antarctic it was considered definite proof of the theory of continental drift.

As well as being a therapsid (mammal-like) reptile, *Lystrosaurus* belonged to the dicynodonts, a group that also contained the small burrow-dwelling **Diictodon** and the water-loving **Placerias**. The dicynodonts were very successful and dominated the landscapes of both the late Permian and early Triassic periods. *Lystrosaurus* was one of the last really successful dicynodont species. As the Triassic progressed, dicynodonts became less common until, by 210 million years ago, they became extinct.

Fascinating Fact > The fossil record suggests that at one time *Lystrosaurus* represented around half of all the large animals on Earth.

Top > A *Lystrosaurus* takes the plunge. It is thought that since they lived and fed near water they were strong swimmers.
Right > A **Proterosuchus** attacks a *Lystrosaurus* as it attempts to make a river crossing. These animals were among the largest land animals of their time, with the extinction event at the end of the Permian period having rendered almost all the world's reptiles extinct.

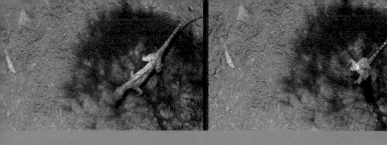
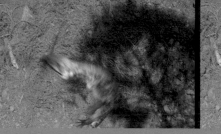

Euparkeria

A possible ancestor to the dinosaurs

name	*Euparkeria* (YOO-park-air-ree-ah), meaning 'early Parker's animal'		**size**	55 cm (22 in) long
animal type	Basal archosaur reptile		**diet**	Carnivorous
lived	245–234 million years ago		**fossil finds**	Africa

Euparkeria was a small and nimble reptile that lived in open woodland areas. It had a light, lean body, long tail, and a small skull with tiny, needle-like teeth. It fed on insects and any other small animals that it could find on the forest floor, and would periodically shed its teeth in order to keep them sharp.

Euparkeria was one of the smallest reptiles of its time, with the adults reaching the size of a large lizard. It lived in a world with many predators, so had to be quick on its feet. It walked on four legs for most of the time, but if a quick getaway was needed, could rise on to its hind legs and run at very high speed. As far as is known, this technique was unique to *Euparkeria* at that time, and would have given it a great advantage. Some people even think that it could have run fast enough to skip lightly across the water surface of small ponds and lakes, just like the present-day basilisk lizard. The only other means of defence that *Euparkeria* possessed was a sharp claw on its thumb, which could have been used as a weapon in close combat.

The first fossils of *Euparkeria* were found in South Africa in 1913, but better specimens were found in 1924. These suggest that the eyesight and sense of smell in *Euparkeria* were excellent, allowing it to find insects, millipedes and other arthropods, and then to pick them up expertly using its jaws. Although probably a solitary animal, it is possible that it may have lived in small groups. The females were larger and more robust than the males, and probably took the lead when it came to mating.

Euparkeria was a primitive member of the archosaurs, a group of reptiles that includes the crocodiles, pterosaurs and dinosaurs (see **Proterosuchus**). Information about the fossil origins of the dinosaurs, pterosaurs and crocodiles is scarce as few fossil reptiles

Fascinating Fact > *Euparkeria* was one of the first reptiles to be able to run on two legs.

are known from the Early Triassic. *Euparkeria* is regarded by some scientists as a possible ancestor to the dinosaurs, although others think that it is more primitive than this. The matter is further confused because there is a gap of 10 million years between *Euparkeria* and the first fossilized dinosaurs, such as **Coelophysis**, and few fossil reptiles are known from this period.

Above > Although we think of prehistoric reptiles as being fearsome killers, a good many (including *Euparkeria*) lived by hunting small insects such as this dragonfly. Just as today, insects would have been abundant but their delicate bodies mean that they are only rarely preserved as fossils.

Top > *Euparkeria* jumps to catch an insect; it was a versatile animal whose athletic capabilities would help lay the foundations for the reptiles' success in the Mesozoic era.

Nothosaurus
A reptile that returned to the sea

name	Nothosaurus (no-thoh-saw-rus), meaning 'false reptile'		size	4 m (13 ft) long
animal type	Nothosaur reptile		diet	Carnivorous
lived	240–210 million years ago		fossil finds	Europe, Middle East, Asia and North Africa

Nothosaurus was one of the few Late Triassic reptiles that could live both in and out of the water. It was medium-sized, had a streamlined body and a long, flattened tail. It hunted in shallow coastal seas, but would come on shore periodically to rest and sun-bathe, as well as to lay its eggs.

When swimming, *Nothosaurus* would use its tail, legs and webbed feet to propel and steer it through the water. The skull was broad and flat, with dozens of needle-sharp, interlocking teeth – perfect for catching and holding on to live fish. *Nothosaurus* hunted by sneaking up slowly on prey, such as shoals of small fish, then putting on a last-minute burst of speed. Once caught, few animals would be able to shake themselves free from the mouth of *Nothosaurus*.

Mating probably took place in the sea. Afterwards the female *Nothosaurus* would go on shore and use her fins to drag herself up the beach. Once above the high-water mark, she would dig a hole and, like modern sea turtles, lay her eggs and then bury them. After hatching, the juveniles would dig themselves free and make a hazardous journey down to the sea, dodging dinosaur predators. Analysis of *Nothosaurus* bones suggests that it became adult at around three years, and lived to six years of age.

Fascinating Fact > *Nothosaurus* may have been a keen sunbather, hauling itself on to the land so that it could bask in the sun's warmth.

The first *Nothosaurus* fossils were discovered in 1834 in Germany, but some exquisitely preserved specimens have recently been recovered from rocks in the Italian Alps. These have allowed scientists to gain a better insight into the lives of these animals.

Nothosaurus belongs to the nothosaurs, a group of marine reptiles that first evolved around 250 million years ago. After the great exinction at the end of the Permian period, the nothosaurs were among the first reptiles to move from the land and into the seas. They would have been a common sight throughout most of the Triassic period, but around 210 million years ago they were joined by many newer, faster marine reptiles, such as the ichthyosaurs, and their numbers began to decline. It is thought that one branch of the nothosaurs evolved into the plesiosaurs, a successful group of fully marine reptiles that includes giant predators, such as **Liopleurodon** and the long-necked **Cryptoclidus**.

Cymbospondylus
The largest ichthyosaur of all time

name	Cymbospondylus (sim-bow-spond-ee-lus), meaning 'boat vertebrae'		size	10 m (33 ft) long
animal type	Ichthyosaur reptile		diet	Carnivorous
lived	240–210 million years ago		fossil finds	North America, Europe and possibly Asia

Cymbospondylus was a gigantic marine reptile that had a large body, eel-like tail and a huge head with a long, sharp-pointed snout. It lived its entire life in water and was by far the largest animal in the sea during the Late Triassic period.

Despite its size, *Cymbospondylus* was not much of a threat to other marine reptiles, such as **Nothosaurus**. Its large jaws contained rows of teeth which were so small that they could not have grasped and held on to large animals, let alone kill them. Instead, the teeth were better designed for catching and holding on to small and medium-sized fish, **ammonites** and belemnites (small squid-like animals). The long tail would have been excellent for swimming, and allowed *Cymbospondylus* to move at fast speeds and efficiently hunt down shoals of swimming fish.

Adult *Cymbospondylus* probably spent much of their time hunting in deep offshore water, only venturing into shallower water to breed or to catch seasonally available prey. Like other ichthyosaurs, *Cymbospondylus* probably gave birth to live young. These, on reaching adult size, probably had few, if any, predators that could harm them.

The first scrappy fossils of *Cymbospondylus* were found in Nevada in 1868 and were later named by palaeontologist Joseph Leidy. It was not until the early 1900s that the first complete skeletons were discovered. Fossil vertebrae from *Cymbospondylus* were allegedly used as dinner plates by Nevada's silver miners; it is now the state's official fossil. While never being common in the fossil record, *Cymbospondylus* managed to survive for much of the Middle Triassic before becoming extinct around 210 million years ago.

Cymbospondylus was the largest and among the oldest known of the ichthyosaurs, a group of exclusively sea-living reptiles that includes dolphin-like hunters, such as **Opthalmosaurus**. Like many of the reptile groups from the Triassic period, the origins of the ichthyosaurs is shrouded in mystery, although it seems likely that they shared a distant common ancestor with the nothosaurs (see **Nothosaurus**). The ichthyosaurs first evolved around 240 million years ago, and went on to become one of the most successful marine groups of the Jurassic period, before they too became extinct around 90 million years ago, in the Cretaceous period.

Fascinating Fact > Although *Cymbospondylus* had a skull that was 1 m (3.3 ft) long, its teeth were small and delicate.

65

Tanystropheus
A reptile that almost defied physics

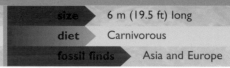

name	Tanystropheus (TAN-ee-STRO-fee-us), meaning 'long vertebrae'		size	6 m (19.5 ft) long
animal type	Prolacertiform reptile		diet	Carnivorous
lived	235–215 million years ago		fossil finds	Asia and Europe

Tanystropheus was about as long as the Nile crocodile but almost three-quarters of this length was taken up by its neck and tail, which makes it look almost impossibly unbalanced. *Tanystropheus* could also live both in and out of the sea, which must surely make it one of the strangest reptiles of all time.

When first discovered, it was thought that *Tanystropheus* would stand on the seashore with its long neck raised in the air, ready to strike down at any fish swimming in the sea below. However, it is now known that the neck was very rigid and contained only 9–12 vertebrae: it couldn't bend up and down, although it did have some side-to-side movement. When *Tanystropheus* swam or walked, the neck would have been held straight out in front of the body, placing great strain on its neck and shoulder region, especially when out of the water. Consequently, *Tanystropheus* must have spent much of its time in the sea, where the weight of the neck and tail would have been supported by the water.

Unsurprisingly, *Tanystropheus* was not a good swimmer. By swishing its tail from side to side, it could have

swum through the water, but only at low speed. Instead of swimming, *Tanystropheus* would probably have spent much of its time standing on or walking along the seabed. Here it would have hunted fish by ambushing them.

With the head so far from the body, *Tanystropheus* could get its mouth close to shoals of fish without alarming them. A quick sideways movement of the neck would then take the fish by surprise, allowing *Tanystropheus* to grab one in its mouth. If *Tanystropheus* hunted on land, it must have picked on animals that could not run away easily, such as insects or small reptiles.

Juveniles did not have a long neck, but gradually acquired one with age. Engineers estimate that for its size *Tanystropheus* has the longest neck length permissible under the laws of physics.

Tanystropheus was discovered in Germany in the 1830s. It was a prolacertiform reptile, which makes it a distant relation of the archosaur reptiles – crocodiles, pterosaurs and dinosaurs. The prolacertiforms reached their peak in the Triassic period and went extinct around 205 million years ago.

Plateosaurus
The first of the dinosaur giants

name	Plateosaurus (PLAT-ee-oh-SAW-rus), meaning 'flat reptile'
animal type	Prosauropod dinosaur
lived	221–210 million years ago

size	8 m (26 ft) long
diet	Herbivorous
fossil finds	Europe

Plateosaurus was one of the first dinosaurs to reach gigantic proportions. It grew to the size of a London bus, but was a vegetarian and lived in the seasonally dry world of the Late Triassic period.

Plateosaurus was perfectly designed for a life spent browsing among trees. The body was wide and round, and held a large stomach and gut that could cope with large volumes of vegetation. The neck was long and sinuous, for reaching high above the ground, and the head long and narrow, with dozens of serrated teeth that could strip leaves and young shoots from trees. *Plateosaurus* would walk mostly on four feet, but it could rise on to two legs when feeding. This allowed it to reach into the treetops, and also meant that its front limbs could be used to pull high branches towards its mouth.

The sense of smell in *Plateosaurus* was excellent and it is thought that the nose had a special scent gland that would have been used to release hormones into the air. These hormones could have been carried many kilometres, and may have been used to attract a mate. *Plateosaurus* probably lived in small herds, and may have migrated for long distances, following the seasonal rains and the lush vegetation they would bring.

Fossils of *Plateosaurus* were first found in Germany in 1837. Since then their remains have been found in vast numbers, suggesting that these dinosaurs were once a common sight.

Plateosaurus belongs to the prosauropods, a primitive group of mostly vegetarian dinosaurs that lived in the Late Triassic and Early Jurassic periods. A prosauropod that lived on Madagascar around 240 million years ago is currently the oldest-known dinosaur on Earth. The relationship between the prosauropods and the rest of the dinosaurs is problematic. It used to be thought that the prosauropods were the direct ancestors of the gigantic sauropod dinosaurs of the Jurassic period, such as ***Diplodocus***, hence the name. But it now seems more likely that these two groups of dinosaurs were cousins and that they shared an as yet undiscovered common ancestor. The prosauropods became extinct around 185 million years ago.

Fascinating Fact > *Plateosaurus* could graze at about the same height as a giraffe.

Placerias
The last of a great dynasty

name	Placerias (pluh-SEHR-ree-us)	size	3.5 m (11.5 ft) long
animal type	Therapsid (dicynodont) reptile	diet	Herbivorous
lived	221–210 million years ago	fossil finds	North America

Placerias was an ox-sized reptile with a powerful neck, strong legs and a barrel-shaped body. It was herbivorous, and would have needed to spend most of its life eating in order to fuel its body. *Placerias* had a beaked mouth with a sharp hook to it that could slice through thick branches and roots. During the wet season, *Placerias* would have eaten fresh leaves, young shoots and moss. In the dry season, when green foliage was rare, *Placerias* would have used its feet to dig in the ground for buried roots and tubers, which its sharp beak would then slice up. Male *Placerias* had especially long tusks that could be used for defence and for frightening rivals during mating displays.

The best *Placerias* fossils come from Arizona, where one location has produced the skeletons of over 40 animals, all of which seem to have died during a severe drought. As the fossils are commonly associated with rivers and lakes, it is surmised that *Placerias* lived a little like the modern hippopotamus. In the wet season it might have spent much of its time wallowing in the water, chewing at bankside vegetation. Remaining in the water would also have given *Placerias* some protection against land-based predators, such as **Postosuchus**.

Placerias was a therapsid (mammal-like) reptile, and is part of the dicynodonts, a group that includes **Lystrosaurus** and **Diictodon**. However, despite the success of the dicynodonts, they became rare during the Late Triassic, possibly because of competition from vegetarian dinosaurs such as **Plateosaurus**. *Placerias* was one of the last of this ancient group of reptiles; the dicynodonts went extinct shortly afterwards.

Fascinating Fact > Fossilized wear-patterns on *Placerias* tusks prove that the animal used them to dig in the ground.

Thrinaxodon
The missing link between mammals and reptiles

name	Thrinaxodon (thrin-AX-oh-don), meaning 'trident tooth'	size	50 cm (20 in) long
animal type	Therapsid (cynodont) reptile	diet	Carnivorous
lived	248–245 million years ago	fossil finds	Africa and Antarctica

Thrinaxodon was a cat-sized animal with a long, low body and a short tail. Although a reptile, it had many mammal-like characteristics, such as whiskers, canines and molar teeth, and probably also milk glands and fur. It lived close to the South Pole and may have been warm-blooded, helping it to cope with cold winters. Even so, it still had a reptilian skeleton and laid eggs.

It is thought that *Thrinaxodon* lived in shallow burrows dug into hillsides or riverbanks. It lived in mated pairs or small family groups, and was probably territorial, using scent glands to mark out boundaries and then defending that territory from intruders. *Thrinaxodon* would sleep in its burrow during daylight hours, emerging at dusk in order to hunt for insects, reptiles and other small prey. Its eyesight and hearing were very good, and the small, stout jaws and front paws would have been deft at finding food in bushes, crevices and under rocks.

There were many large predators during the Late Triassic, including some of the earliest carnivorous dinosaurs, such as **Coelophysis**. *Thrinaxodon* had few defences against these, and its main survival strategy would have been to feed at night and to sense, then hide from approaching animals.

As a reptile, *Thrinaxodon* was an egg-layer, but it is thought likely that the females had milk-producing mammary glands so that, after hatching, the juveniles could suckle on their mothers. This would have provided the young with an enormous advantage over other reptiles, whose parents would have had to leave them in the nest or burrow in order to forage for food. Mammary glands greatly increased the chances of a hatchling's survival, although it did mean that the parents could produce only a few young at a time (as opposed to the dozens produced by some reptiles) because there is only so much milk a mother can supply.

Thrinaxodon fossils were discovered first in South Africa, then later in Antarctica, supporting the idea that the two continents were once joined together.

Fascinating Fact > *Thrinaxodon* had long whiskers, like a cat.

Thrinaxodon was a cynodont, a subgroup within the therapsid (mammal-like) reptiles (see **Gorgonops**). The cynodonts were very advanced reptiles, and in many respects were only a short evolutionary leap away from being true mammals. *Thrinaxodon* had so many mammalian features that it is considered by many scientists to be a link between the mammals and the reptiles.

Above > This near-perfectly preserved *Thrinaxodon* skull shows its specialized teeth and articulated lower jaw – features that allowed *Thrinaxodon* and other cynodonts to make the most of their food.

Coelophysis
The dinosaurs' small beginnings

name	Coelophysis (SEE-low-FY-sis), meaning 'hollow form'	size	3 m (10 ft) long
animal type	Theropod (coelurosaur) dinosaur	diet	Carnivorous
lived	227–210 million years ago	fossil finds	North America

Coelophysis is one of the oldest-known meat-eating (theropod) dinosaurs. It was small, could walk on two legs, and had a long neck and small head. Its jaws were lined with dozens of needle-like teeth. *Coelophysis* bones were partially hollow and very light; this allowed it to be nimble, and to run fast and jump, something that few other animals its size could do. It also had small arms with strong, clawed fingers, which could have been used to grasp prey, or to scavenge on the ground when *Coelophysis* dropped to all fours. The long, slender tail acted as a counterbalance, helping it to stay upright and change direction suddenly, an advantage when running at speed.

The brain of *Coelophysis* was large in comparison to that of its reptilian ancestors, and its senses were finely tuned. It had excellent eyesight and hearing that, in combination with the fast legs, long neck, short arms and sharp claws, would have helped it catch fish, small reptiles, amphibians, insects and other prey.

The environment in which *Coelophysis* lived was subject to long droughts, when many individuals would die of thirst and starvation. During these harsh times food was scarce, and *Coelophysis* would sometimes resort to cannibalism, attacking and eating its own juveniles.

Like modern wolves, *Coelophysis* may have lived and hunted in packs. The females were slightly smaller than the males, but there is no evidence that males fought one another for mating rights. Newly hatched *Coelophysis* would have had to fend for themselves, but they would have grown quickly and reached full size by around seven years.

Top right > Hundreds of *Coelophysis* fossils have been found at Ghost Ranch quarry in the USA.
Below > Cannibalism is a practicable solution to starvation.

Fascinating Fact > Adult fossil *Coelophysis* have been found with the bones of juvenile *Coelophysis* in their stomachs, suggesting that they went in for cannibalism.

The first *Coelophysis* fossil was discovered in 1881 in New Mexico by palaeontologist David Baldwin, but the skeleton was very incomplete. Little was known about this small dinosaur until 1947, when literally hundreds of complete *Coelophysis* skeletons were discovered in Ghost Ranch quarry, also in New Mexico. This find is believed to be a herd of *Coelophysis* killed either by drought or a sudden flash flood.

Some of the adult skeletons were found to have juvenile skeletons within them. At first it was thought that these were unborn *Coelophysis* (which would have meant that the animals gave birth to live young, like most mammals), but further study revealed that the juveniles had in fact been eaten by the adults.

Coelophysis is part of the theropod group of dinosaurs, which in only a few million years evolved to produce gigantic predators, such as ***Allosaurus*** and ***Giganotosaurus***.

The origin of the dinosaurs, however, remains a mystery. Scientists know that they evolved from the basal archosaurs (sometimes called 'thecodonts'), a group of reptiles that included ***Euparkeria***, and lived in the early Triassic period. Recent discoveries in Brazil have pushed the age of the oldest-known theropod dinosaur fossils to around 235 million years, but there is a gap in the fossil record of around 10 million years between this and any probable dinosaurian ancestors.

Peteinosaurus
A delicate flying reptile

name	*Peteinosaurus* (pet-INE-oh-SAW-rus), meaning 'winged lizard'	size	60-cm (2-ft) wingspan
animal type	Pterosaur reptile	diet	Insectivorous
lived	221–210 million years ago	fossil finds	Europe

Peteinosaurus is one of the oldest-known pterosaurs, but it is tiny when compared to some later species, such as **Pteranodon**. Like most pterosaurs, *Peteinosaurus* had bones that were strong but very light, allowing it to have large wings but a very low body weight. *Peteinosaurus* and other pterosaurs had wings that were made of skin stretched between its arms and legs. The wings had a large surface area, but weighed almost nothing (unlike those of birds, which are covered in heavy feathers). It was this combination of large overall size but small body weight that made the pterosaurs such accomplished fliers.

The oldest pterosaurs, including *Peteinosaurus*, had long, stiff tails that helped the animal make precise movements in the air. *Peteinosaurus* used its flying skills to hunt on the wing, snatching insects out of the air using its short beak and sharp, conical teeth.

Below > To prevent overheating in the midday sun, some small pterosaurs cooled off by dipping their heads under water.

Before the pterosaurs evolved, there had been numerous other types of flying reptiles, but they were capable only of gliding between trees (a bit like present-day flying squirrels). *Peteinosaurus* and other pterosaurs were different: they could flap their wings. This meant that they could take off from the ground and keep themselves in the air for as long as they needed. Their large wings also meant that they could glide on the air current and thus save energy.

The first fossils of *Peteinosaurus* were found in 1978 in the Italian Alps by palaeontologist Rupert Wild. The discovery caused great excitement at the time because the fossils belonged to the oldest-known pterosaur and were extremely well preserved. It is possible that *Peteinosaurus* was ancestral to some of the better-known Jurassic pterosaurs, such as **Rhamphorhynchus**. *Peteinosaurus* is still known from only two incomplete skeletons.

The pterosaurs were archosaurs, a broad group of reptiles that also includes the dinosaurs and crocodiles. The origin of all the archosaurs is problematic. There are many species of basal archosaur (sometimes called thecodonts, e.g. **Euparkeria**) at the start of the Triassic period, but there then follows a gap of 10 million years, where few archosaur fossils are found. It is from around 235 million years ago that the first fossilized pterosaurs, dinosaurs and crocodilians are found, but tracing their evolutionary history prior to this has proved troublesome. In particular, pterosaurs suddenly appear in the fossil record as highly specialized fliers with no clear intermediates before them.

It is generally agreed that the pterosaurs probably evolved from small, lizard-like creatures that would have habitually climbed or lived in trees. These reptiles would, in time, have developed the ability to glide between trees, using the thin flaps of skin that ran between their arms and legs. A fully fledged ability to fly evolved from these humble beginnings. There are a number of gliding reptiles known from the Late Permian and Early Triassic periods (256–242 million years ago), but so far none is thought to have been an ancestor to the pterosaurs.

Fascinating Fact > Despite having a 60-cm (2-ft) wingspan, *Peteinosaurus* probably weighed less than a blackbird.

Postosuchus
A hunter of dinosaurs

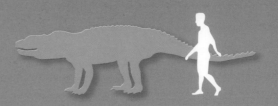

name	*Postosuchus* (POHST-oh-SOOK-uss), meaning 'after crocodile'	size	5 m (16.5 ft) long
animal type	Archosaur reptile	diet	Carnivorous
lived	227–210 million years ago	fossil finds	North America

The forests of the Late Triassic period were home to the large and terrifying *Postosuchus*, a carnivorous hunter that could attack and kill almost any animal of its time. Its skull was massive and much taller than it was wide, which, like that of **Tyrannosaurus**, made its bite stronger. Its mouth was lined with sharp, serrated teeth up to 8 cm (3 in) long that could cut and rip at the toughest of flesh.

Postosuchus was a well-defended animal. Its back was covered in a layer of thick armoured plates, while its head had a number of solid bony projections. Its front limbs possessed a long, sharp claw that could be used both for defence and attack. It was flat-footed, but its long legs were tucked underneath its body, which meant that it could run at speed: it may also have been able to rise up on to two legs, which would have helped it attack larger animals, such as **Placerias**. It could probably also have attacked the many small dinosaurs, such as **Coelophysis**, that existed at the time.

Top right > With its powerful jaws and sharp teeth, the large reptile *Postosuchus* was one of the first super-predators to evolve in the Mesozoic era

Below > *Postosuchus* was probably a lone predator and would have developed an aggressive display towards competing animals.

As top predator, *Postosuchus* probably lived a solitary life, avoiding other members of its species (apart from when mating). It would have hunted other large animals by a combination of stealth and ambush so as to take its prey by surprise.

Several *Postosuchus* fossils have been found in Texas and Arizona. These reveal that the animal was an archosaur reptile, and thus part of the same broad group that contained the dinosaurs and pterosaurs, although it was only distantly related to these. *Postosuchus* is generally thought to be a very primitive type of crocodilian (the term 'crocodile' refers only to the modern species).

The origins of the crocodilian reptiles are still shrouded in mystery. They split from the primitive archosaurs in the Middle Triassic, around 240 million years ago, but the first true crocodilians occurred around 35 million years later. It would appear that *Postosuchus* lay somewhere along that line of evolution. The true crocodilians went on to be hugely successful, producing numerous large and small species, including sea-going ones, such as **Metriorhynchus**, and the gigantic **Sarcosuchus**. The crocodilians eventually also gave rise to the crocodiles, alligators and gavials that are with us today.

Ammonites
Fossil icons

name	Ammonite (am-on-ite), named after the Egyptian god Ammon	size	2.5 cm – 2.5 m (1 in – 8 ft) in diameter
animal type	Cephalopod mollusc	diet	Carnivorous
lived	400–65 million years ago	fossil finds	Worldwide

The ammonites were shellfish that could be found in all the seas and oceans around the world. There were many hundreds of ammonite species, and they adopted a wide variety of shapes and lifestyles. However, the one thing they all had in common was a coiled shell, the inside of which was subdivided into several chambers. Only the outermost chamber was occupied by the animal itself; the remaining chambers could be filled with either water, to help it sink, or air, to help it float. By adjusting the balance of water to air in its shell the animal could remain buoyant in the water.

The ammonite animal itself lived almost entirely inside its shell; only its tentacles, mouth, eyes and hyponome (a tube through which it squirted water to move it through the water) would have been visible.

Ammonites were hunters, but they were not fast swimmers, and spent much of their time 'hanging' in the water or bobbing along at a slow speed. This limited the sort of food that they could catch, and meant that most ammonites lived on slow-moving crustaceans, such as crabs and lobsters, or floating animals, such as jellyfish. Such animals would be captured by the tentacles, then fed to the mouth, which consisted of a pair of jaws that together formed a sharp beak.

The shape and patterning of ammonite shells varies greatly. Some species were small and plain-looking, while others were massive and highly ornate, their shells covered with whorls, ridges, lumps, bumps and intricate patterns. One group of ammonites (the heteromorphids) even had shells that were fully or partially uncoiled, making them look like twisted bits of wire.

Throughout the fossil record there is a very high turnover of ammonite species. This variety, plus the visible differences between the shells, has been used to good effect by geologists. Since each species of ammonite is found only at a certain period in time, it is possible for geologists to gauge roughly how old a particular rock is, based on the ammonite species that

Top > Ammonites were abundant in the world's oceans throughout much of the Mesozoic era but they were an easy source of food for many animals. When confronted by an approaching predator, such as an *Ophthalmosaurus*, there was little that the ammonite could do to avoid being eaten except retract into its shell.

Left > The ammonites went extinct over 65 million years ago but their cousins, the nautiloids, have survived through to the modern day. It is from studying living nautiloids, such as the two *Nautilus* shown here, that we have been able to understand much about the biology and behaviour of the extinct ammonites.

Right > The abundance of ammonite fossils in many Mesozoic rocks has long made them favourite targets for amateur fossil collectors. Their classic coiled shape has also made ammonites one of the most instantly recognizable of all fossils.

occur within it. Prior to the development of high-technology dating techniques (such as radioactive dating), ammonites were one of the few reliable means of estimating the age of particular rock outcrops.

Being slow-moving and abundant, the ammonites had many predators and were the staple diet of many animals. Tooth-marks in their shells show that they were eaten by sharks, plesiosaurs, ichthyosaurs, mosasaurs and myriad other large swimming marine life. Ammonites had few (if any) means of protection, although some species did develop strengthening ridges and small spines on their shell, perhaps to deter certain predators from biting them. Even if the bite of a predator managed to miss the main body of the animal, a serious puncture in the shell could still kill an ammonite, as it would be unable to control its buoyancy in the water and simply sink to the bottom.

Fossils of ammonites are abundant and can be found in many places around the world. Sometimes their fossils are so plentiful that they erode out of the rocks and litter large stretches of seashore or riverbed. The classic spiral ammonite shape has been familiar to generations of people. Legend has it that in medieval times people used to think that ammonites were snakes that had been turned to stone as a punishment from God. In order to play up to this legend, some enterprising stonemasons would sometimes carve a snake's head on to an ammonite fossil. Nowadays ammonites are so familiar to us that their image has become a media icon used in everything from television advertisements to company logos.

The ammonites were cephalopod molluscs, which means that they are in the same broad group as squid, octopuses, cuttlefish, the present-day nautilus and orthocones, such as *Cameraceras*. The earliest fossil ammonoids (the subdivision of cephalopods to which ammonites belong) are found way back in the Devonian period, where they probably evolved from a group of straight-shelled cephalopods known as the bactritoids. The ammonoids were of relatively little importance in the Palaeozoic seas (with the possible exception of a subgroup known as the goniatites, which have proved useful to geologists), and they almost died out during the great extinction event that occurred at the end of the Permian 248 million years ago. Those few ammonoid species that did survive into the Mesozoic era quickly multiplied, filling the world's seas with dozens of ammonite species. They continued to be abundant and widespread throughout the Triassic, Jurassic and Cretaceous periods, occurring everywhere from the Arctic Ocean to the coasts around Antarctica. However, towards the very end of the Cretaceous the ammonites began to decline in number, with the last-known fossils being found just prior to the great extinction that wiped out the dinosaurs, 65 million years ago.

Leedsichthys
The biggest fish ever

name	Leedsichthys (leeds-ICK-thees), meaning 'Leeds's fish'
animal type	Teleost fish
lived	165–155 million years ago

size	Up to 27 m (88.5 ft) long
diet	Carnivorous
fossil finds	England, France and Chile

At around the size of a blue whale, *Leedsichthys* was the largest-known fish of all time, but, although its size was daunting, it was a gentle giant with a simple lifestyle. *Leedsichthys* swam near the sea's surface, taking in mouthfuls of water rich in plankton, tiny shrimps, jellyfish and small fish, which it sieved through the mesh plates at the back of its mouth. These plates were made from tens of thousands of fine teeth.

To find enough food to eat, it is thought that *Leedsichthys* would travel to those parts of the world where nutrient-rich, deep-sea currents would rise to the ocean surface. The excess of nutrients in these currents would feed the plankton, which would become so plentiful that it formed a dense organic soup. Such an abundance of plankton would attract shoals of *Leedsichthys* (and other filter-feeding fish) intent on gorging themselves. Plankton feasts such as these could last for weeks, after which a large animal like *Leedsichthys* would be able to survive for several more weeks without feeding. During this time it would take the opportunity to shed the giant filter plates from the back of its mouth and wait while new ones grew back.

The Jurassic seas were dangerous, and despite its gigantic size *Leedsichthys* had no means of defending itself against large and persistent attackers. Fossilized tooth-marks in its skeleton show that it would have been attacked by the marine crocodilian **Metriorhynchus**, but doubtless other large predators, such as the giant pliosaur **Liopleurodon**, would have had a go too. Disabling a *Leedsichthys* by attacking its fragile fins or tail would have been an easy task for a predator; killing it would have taken much longer. An injured *Leedsichthys* might have taken days to die and in the meantime would have been slowly eaten alive by its attackers.

The first *Leedsichthys* bones were discovered in the late 1880s near to Oxford, England, by Alfred Leeds, a farmer and keen amateur palaeontologist. The fish is known from only a few isolated finds, although a reasonably complete specimen was recently found stored in a cupboard in a Glasgow museum.

Fascinating Fact > *Leedsichthys* is thought to have had over 40,000 teeth.

Leedsichthys was a teleost, which means that it belongs to the same group as most present-day fish. The teleosts first evolved around 240 million years ago and are a subgroup within the more ancient 'bony fish' (the actinopterygii or ray-finned fish), whose origins go all the way back to the Early Devonian period (about 410 million years ago). *Leedsichthys* was a primitive example of a teleost fish, but the group as a whole soon expanded rapidly, and by the end of the Cretaceous period (65 million years ago) they had started to dominate the seas.

In the modern oceans the teleost fish are by far the most widespread and abundant, with an estimated 22,000 living species (the next biggest group of living fish is the sharks and rays, see **Hybodus**).

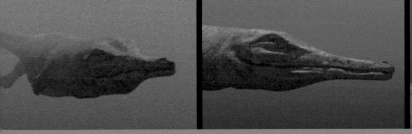

Metriorhynchus
An ocean-going crocodile

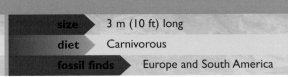

name	Metriorhynchus (MET-ri-oh-RINK-us), meaning 'moderate snout'	size	3 m (10 ft) long
animal type	Crocodilian (metriorhynchid) reptile	diet	Carnivorous
lived	160–150 million years ago	fossil finds	Europe and South America

The marine crocodilian *Metriorhynchus* was a sleek animal that spent almost its entire life at sea. Its body was streamlined, with few of the scaly lumps and bumps that modern crocodiles possess. Its tail was long and powerful, and would have propelled it gracefully through the water by the use of a strong, sideways-sweeping motion.

The fossils of *Metriorhynchus* were first identified by the French scientist Henri de Blainville in 1833. Although initially known only from the Jurassic rocks of England and France, *Metriorhynchus* fossils have now also been found in Chile and Argentina.

Fossilized stomach contents show that *Metriorhynchus* was a very versatile predator that would hunt everything from slow-moving ammonites to faster-moving large fish. It is known to have tackled the giant fish **Leedsichthys** and also to have successfully caught flying pterosaurs (such as **Rhamphorhynchus**), which it could have grabbed in mid-air by leaping out of the water. Despite its powerful hunting ability, *Metriorhynchus* was relatively defenceless against larger hunting reptiles and may have fallen victim to predatory marine giants, such as **Liopleurodon**.

Metriorhynchus was so adapted to life at sea that it returned to land only to mate and lay its eggs. It was not very graceful when out of the water: its legs were better adapted to swimming than walking, and moving across the beach would have involved dragging and pushing its body with its limbs, a bit like a large turtle.

Top > *Metriorhynchus* was a silent predator of the world's oceans. Here an individual prepares to attack a large fish by mounting an assault from the rear.
Left > A *Metriorhynchus* and a **Hybodus** shark launch a joint attack on the giant fish **Leedsichthys**. Fossils show that such attacks did occur and that there was probably little that **Leedsichthys** could do to prevent them. *Metriorhynchus* was an opportunistic predator that would eat almost anything that strayed too close to its mouth.

Fascinating Fact > Instead of having feet and claws, *Metriorhynchus* had paddles.

After laying its eggs, *Metriorhynchus* would have returned to the sea immediately. The young probably hatched on their own, then made a hazardous journey down the beach to the sea, where they would have faced dangers from various predators, including dinosaurs and flying pterosaurs.

Metriorhynchus belongs to a group of marine crocodilians known as the metriorhynchids, of which there were several species living in the seas of the Late Jurassic period. (It is not strictly correct to refer to *Metriorhynchus* as a crocodile; that is a term reserved only for modern crocodiles and alligators, which evolved much later in time; see **Sarcosuchus**.) The Jurassic period was a very good time for the crocodilians, and metriorhynchids were just one of many different types of crocodilian that were then alive.

The crocodilians are archosaurs, the reptile group that contains the dinosaurs and pterosaurs. Like them, the crocodilians have obscure origins, but are believed to have evolved around 225 million years ago from a basal archosaur animal, such as **Proterosuchus**. The earliest confirmed crocodilians are found in the Early Jurassic period (206–180 million years ago), and within a short period of time the crocodilians could be found living on land and in rivers, lakes, estuaries and seas across the globe. Their versatility made them hugely successful, but most species were small fish-eaters, being only 1–3 m (3.3–10 ft) in length (under half the length of the largest living crocodiles). After the end of the Jurassic the crocodilians continued to evolve, growing in size and stature until they produced some truly gigantic species (see **Sarcosuchus**), as well as the ancestors to the true crocodiles (the crocodylia) that still exist today.

Rhamphorhynchus
A Jurassic 'seagull'

name	*Rhamphorhynchus* (RAM-for-INK-us), meaning 'beak nose'		size	1.8-metre (6-ft) wingspan
animal type	Pterosaur reptile		diet	Carnivorous
lived	170–144 million years ago		fossil finds	Europe and Africa

Rhamphorhynchus was about the same size as an albatross, yet is still one of the smallest-known pterosaurs. It lived close to the sea, where it would hunt fish by swooping low over the water and momentarily dipping its beak in to scoop out a fish. Its forward-facing, sharp, conical teeth ensured that the fish were firmly gripped and couldn't slip out again once caught. Some *Rhamphorhynchus* fossils suggest that it had a throat pouch in which it could store spare fish before returning to land to digest them.

There are several known species of *Rhamphorhynchus*, the smallest of which had a wingspan of only 40 cm (15 in). All are noted for their distinctive long skulls, with sparse but sharp teeth, and long tails. They are generally considered to have been accomplished fliers, and would have manoeuvred skilfully, possibly even crossing large bodies of water by using the lift from air currents above the waves.

Fossils of *Rhamphorhynchus* are relatively common and have been found in England, Germany, Portugal and Tanzania. Undoubtedly the best-preserved specimens are those found in the Solnhofen limestone of Bavaria, Germany (the same quarry that produced the first fossil bird *Archaeopteryx*). The Solnhofen fossils of

Below > *Rhamphorhynchus*'s bizarre teeth were perfect for grasping slippery fish.

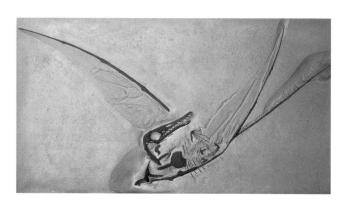

Above > This fossil *Rhamphorhynchus* is so well preserved that delicate tissues such as the wings are clearly visible.

Rhamphorhynchus are justifiably famous; not only are they beautiful, but they also preserve even the softest parts of the body, such as the delicate skin of its wings and the stomach contents. Indeed, some of the fossils are so well preserved that they suggest *Rhamphorhynchus* might have had a layer of hair covering its body. This remained uncertain until the 1970s, when another related Jurassic pterosaur called *Sordes pilosus* ('hairy devil') was discovered in Russia. This fossil appeared to show that the animal had had a dense layer of hair covering virtually the whole of its body.

Hairy animals are usually considered to have been warm-blooded (the hair being used as insulation against the cold). The presence of hair on pterosaurs (and their warm-bloodedness) remains a controversial topic. Some scientists argue that the hairs seen on the fossils are in fact loose muscle fibres and not real hair at all. However, some of the hair occurs in places where muscles do not exist. The debate as to whether the pterosaurs were hairy and warm-blooded or naked and cold-blooded dates back to the 1860s, and has still to be properly resolved.

Rhamphorhynchus was an archosaur reptile and a distant descendant of Triassic pterosaurs, such as **Peteinosaurus**. Like most Triassic and Jurassic pterosaurs, *Rhamphorhynchus* had a long tail that it used to help balance itself during flight. However, it was one of the last species of the long-tailed pterosaur ever to exist. After the start of the Cretaceous period (around 144 million years ago) the pterosaurs lost their tails and became much larger; some even reached the size of small aeroplanes (see **Ornithocheirus**).

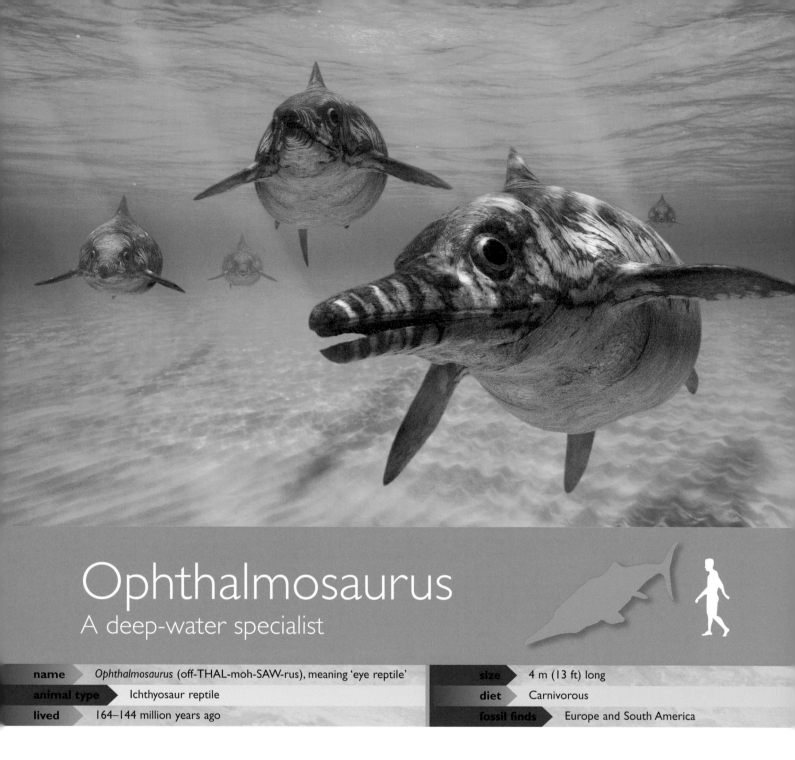

Ophthalmosaurus
A deep-water specialist

name	*Ophthalmosaurus* (off-THAL-moh-SAW-rus), meaning 'eye reptile'	**size**	4 m (13 ft) long
animal type	Ichthyosaur reptile	**diet**	Carnivorous
lived	164–144 million years ago	**fossil finds**	Europe and South America

Ophthalmosaurus was a medium-sized ichthyosaur marine reptile that inhabited the warm tropical waters of the Late Jurassic period. It had a long pointed snout, and its body was streamlined and dolphin-like in shape, which allowed it to swim at high speed in pursuit of fish, squid and other small marine animals.

Its large eyes (after which it is named) gave it excellent eyesight, even in almost pitch-black conditions. This allowed it to make deep dives into darker waters, where competition for food was less than in the crowded surface waters. As an air-breather, *Ophthalmosaurus* would not have been able to live in the deep ocean; it would have needed to return to

the surface regularly in order to breathe. It could, however, have held its breath for extended periods, possibly up to 20 minutes, which would have given it plenty of time to make a deep dive, hunt and then return to the surface.

Like all ichthyosaurs, *Ophthalmosaurus* had a wide, semi-rigid, vertical tail, which it would have beaten rapidly from side to side. Doing this could have propelled it through the water at high speed, and it would

Fascinating Fact > *Ophthalmosaurus* had the largest eyes of any backboned animal.

steer using its front fins. When travelling flat out, it could have moved at an estimated 40 kph (25 mph) or more. This almost certainly made the ichthyosaurs the fastest marine reptiles in the Jurassic seas – too fast in full flight for large predators, such as *Liopleurodon*, to catch. At top speed *Ophthalmosaurus* could have jumped clear of the water to grab a breath without slowing down.

The long, narrow snout of *Ophthalmosaurus* was ideal for picking off small and medium-sized fish. The jaws contained only a few small teeth, which, rather than killing fish outright, were probably used to grip them momentarily before swallowing them whole. Fossilized dung thought to be from *Ophthalmosaurus* suggests that it ate a wide range of animals, from bottom-dwelling fish to other smaller marine reptiles.

Some ichthyosaur fossils are so well preserved that it is possible to see the outline of the skin and the contents of the stomach. It is also possible to see the fossils of tiny embryos inside the body of the female. This means that ichthyosaurs did not lay eggs like most other reptiles, but instead gave birth to live young. Fossils show that these were born tail first so that they did not drown before reaching the surface. Most ichthyosaurs would give birth to only one or two young, but some produced up to 11 in a single go.

The first ichthyosaur fossils were found in Europe during the 1820s, and for many decades they were highly prized, fetching large sums of money for those that found them. *Ophthalmosaurus* itself was not discovered until 1907. Its compact, dolphin-like shape is typical of the Jurassic ichthyosaurs, but differs from the more primitive, eel-like shape of older Triassic ichthyosaurs, such as *Cymbospondylus*. The ichthyosaurs were highly successful marine predators during the Jurassic, and their fossils from that time are common. However, the ichthyosaurs started to become less common during the Cretaceous period, and eventually went extinct around 90 million years ago. Their role as marine predators was taken over by plesiosaurs such as *Elasmosaurus* and large diving birds such as *Hesperornis*.

Above > Ichthyosaurs such as *Ophthalmosaurus* were once a common sight in all the world's oceans. Although large and swift-moving, the narrow snout and small teeth of most ichthyosaurs meant that they were only capable of eating small animals such as ammonites and jellyfish. The lifestyle of many ichthyosaurs is often compared to that of living dolphins, which are about the same size and hunt in a similar manner. Whether or not ichthyosaurs were as playful as dolphins is not known.

Below > This classic fossil from Germany shows a female ichthyosaur that was killed in the process of giving birth. Fully marine reptiles like ichthyosaurs could not come ashore to lay eggs (as turtles do) and so had to give birth to live young. Fossils like this reveal that the newborn ichthyosaur would be delivered into the world tail first; this was so that it wouldn't drown before its mother could nudge it to the surface to take its first breath.

Liopleurodon
The biggest killer of all time

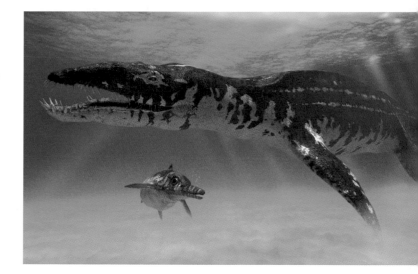

name	Liopleurodon (LIE-PLOO-ro-don), meaning 'smooth-sided tooth'	size	At least 25 m (82 ft) long
animal type	Plesiosaur (pliosaur) reptile	diet	Carnivorous
lived	160–155 million years ago	fossil finds	Europe and possibly Central America

Liopleurodon was the mightiest marine predator of all time. Its 25-m (82-ft) long body would have cruised silently through the shallow seas of the late Jurassic period, propelled by the alternate flapping of its four gigantic flippers. This form of swimming is thought to have been unique to the plesiosaurs and has not been used by any other animal.

The skull of *Liopleurodon* is characteristic of the pliosaurs, being long, heavy and attached to its body by a short neck. The long jaws and rows of needle-sharp teeth were capable of killing any other animal in the seas at the time. The skull and jawbones were specially strengthened to allow them to withstand the powerful biting force produced by the jaws. Marine crocodiles, the giant fish **Leedsichthys**, ichthyosaurs and even other pliosaurs were all capable of being attacked by *Liopleurodon*.

While most marine reptiles had to close their nostrils when swimming, pliosaurs such as *Liopleurodon* had evolved a nose that allowed them to smell while holding their breath under water. Using this height-ened sense, *Liopleurodon* could have smelt its prey from some distance away and, like a modern shark, have followed the scent to its source. *Liopleurodon* had very good eyesight, and once the prey was spotted would put on a quick burst of speed using its enor-mous flippers: the prey would be gone, victim to its powerful jaws.

Despite needing to breathe air, *Liopleurodon* spent its entire life at sea and was too large and heavy to leave the water, even for short periods of time. As a consequence, it would probably have given birth to live young. To do this it might have gone into shallow water, which offered more protection. The juvenile *Liopleurodon* would probably have remained in the shallows until they had grown sufficiently to move further away from shore.

Fascinating Fact > *Liopleurodon*'s mouth was three times bigger than that of **Tyrannosaurus**.

Fossils of *Liopleurodon* were discovered in France in 1873. The longest specimen among these was 18 m (59 ft), but recently another large marine reptile has been found with tooth-marks in it that came from a pliosaur (probably *Liopleurodon*) that was at least 25 m (82 ft) long.

The pliosaurs were members of the plesiosaur order of marine reptiles and *Liopleurodon* was part of a group that is sometimes also called 'short-necked plesiosaurs'. The plesiosaurs are thought to have evolved from the nothosaurs (see **Nothosaurus**) in the late Triassic period (227–206 million years ago). The plesiosaurs were especially common during the Early Jurassic period, when, in addition to the short-necked pliosaurs, there were also long-necked varieties (e.g. **Cryptoclidus**). There were many species of pliosaur living during the Jurassic period, but by the Cretaceous period they were rarer and eventually became extinct around 80 mil-lion years ago. Other long-necked species of plesiosaur continued to survive into Late Cretaceous times (see **Elasmosaurus**).

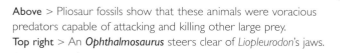

Above > Pliosaur fossils show that these animals were voracious predators capable of attacking and killing other large prey.
Top right > An **Ophthalmosaurus** steers clear of *Liopleurodon*'s jaws.

Cryptoclidus
An elegant fishing reptile

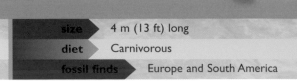

name	Cryptoclidus (KRIP-toe-KLIDE-us), meaning 'hidden collar-bone'	size	4 m (13 ft) long
animal type	Plesiosaur reptile	diet	Carnivorous
lived	164–155 million years ago	fossil finds	Europe and South America

Cryptoclidus was a long-necked plesiosaur that lived in the shallow coastal seas around western Europe. Like all plesiosaurs, *Cryptoclidus* was a predator, but its small head and teeth would have limited its prey to small anchovy-sized fish, **ammonites** and squid. It would have hunted by swimming into large shoals of fish and using its 2-m (6.5-ft) neck to grasp them. Its bulky body meant that it was not a fast swimmer, so instead of pursuing its prey it would have had to ambush it or strike suddenly. It has also been suggested that *Cryptoclidus* could have used its thin teeth to sift through soft seabed sediments in search of crabs, small fish and worms. Once caught, its prey was swallowed whole.

Being one of the smaller plesiosaurs, *Cryptoclidus* could possibly have hauled itself out of the water in order to rest, escape danger or perhaps even to breed. On land, however, it would not have been mobile, and could not have strayed far from the water's edge.

In the water *Cryptoclidus* used the four-flipper swimming action unique to all plesiosaurs. When cruising, it moved by flapping each pair of rhomboidal flippers alternately; if it needed to make a lunging movement, all four flippers would move together, thrusting it forwards in the water. *Cryptoclidus* would have been an elegant sight to watch, appearing to fly under water.

The first *Cryptoclidus* fossils, discovered in England in 1892, showed that the animal was an example of a long-necked plesiosaur (as opposed to a short-necked plesiosaur, such as ***Liopleurodon***) and a distant relative of ***Elasmosaurus***. As a slow-moving and relatively small marine reptile, *Cryptoclidus* may have been preyed upon by larger predatory species, such as ***Liopleurodon***.

Above > Despite their size, long-necked plesiosaurs such as *Cryptoclidus* could not tackle large prey and so were not a threat to most animals. They are sometimes portrayed as being curious and playful animals (a bit like modern seals) who would have frolicked in the shallow seas. They would, however, have been vulnerable to attack by large sharks and pliosaurs, and would have needed to keep their wits about them at all times.

Right > The graceful alternate flapping motion of the plesiosaur's flippers has been described as a type of underwater flying. Scientists were long puzzled by how plesiosaurs could have used these rhomboidal flippers to swim. Even though they would have been efficient and powerful swimmers, the plesiosaurs' technique has not been adopted by any large animals living in today's oceans.

Hybodus
A Jurassic shark

name	*Hybodus* (hie-BOW-dus), meaning 'humped tooth'
animal type	Hybodont shark
lived	230–90 million years ago

size	2 m (6.5 ft) long
diet	Carnivorous
fossil finds	Asia, Europe, Africa and North America

Hybodus was a remarkable shark that for tens of millions of years was very common in the seas around the world. It was not very big, but had the classic streamlined shark shape, complete with two dorsal fins that would have helped it steer with precision. The mouth was not large, and rather than ruthlessly hunt large prey, *Hybodus* was capable of eating a wide range of foods. Its jaws contained two types of teeth: sharp ones for seizing slippery prey, such as fish and squid; flat and strong ones for crushing shelled animals, such as molluscs and sea urchins.

Shark skeletons are made of soft cartilage, so do not fossilize well. However, there are a few rare *Hybodus* fossils showing its spiny dorsal fin, which was used as a means of defence. If another sea creature tried to swallow the shark, *Hybodus* could raise its dorsal spine so that it would stick in the roof of the predator's mouth, choking it to death. It is unlikely that *Hybodus* was a fast swimmer, but as its prey was small and slow-moving, it did not need to be.

The first fossilized teeth from *Hybodus* were found in England around 1845. Since then teeth (and occasionally dorsal spines) have been recovered from around the world. The hybodonts (the order of sharks to which *Hybodus* belongs) are very ancient and probably had their origins in the late Devonian period (around 370 million years ago), when they would have competed for food with other primitive sharks, such as **Stethacanthus**. During the Triassic, Jurassic and Cretaceous periods the hybodonts were especially successful, and could be found in shallow seas across the world. During much of the Jurassic and Cretaceous the hybodonts shared the seas with the neoselachian sharks, the group that includes all present-day species of shark. For reasons that are not fully understood, the hybodonts became extinct near the end of the Cretaceous period. The neoselachian sharks carried on to dominate the oceans with forms such as the megalodon shark (**Carcharodon megalodon**) and its modern relative the great white shark.

Eustreptospondylus
A coastal carnivore

name	*Eustreptospondylus* (yoo-STREP-toe-spon-DIE-lus)		size	5–7 m (16.5–23 ft) long
animal type	Theropod (carnosaur) dinosaur		diet	Carnivorous
lived	164–159 million years ago		fossil finds	Europe

Eustreptospondylus was a medium-sized theropod dinosaur that lived near the coast. Like all theropods, it was bipedal and carnivorous, and may have lived by hunting smaller dinosaurs and pterosaurs, or by scavenging for the dead bodies of fish, dinosaurs and marine reptiles along beaches and estuaries.

There is little firm information about *Eustreptospondylus* because it is known only from a single incomplete juvenile specimen discovered near Oxford, England. Around 160 million years ago, during the Middle Jurassic period, this part of southern England was a series of small islands surrounded by a shallow sea. One suggestion is that *Eustreptospondylus* lived on these islands and that it was capable of swimming between them. This is not as far-fetched as it might sound; many modern large reptiles are capable of swimming considerable distances. The Komodo dragon, for example, has been known to swim between islands in the Java straits. An alternative theory is that the *Eustreptospondylus* whose fossil was found in England lived on the European mainland, and that its body fell into a river or estuary and was then swept far out to sea from the land, perhaps during a flood.

Eustreptospondylus is usually placed in the carnosaur group of dinosaurs. The carnosaurs are an ill-defined collection of theropods, whose evolutionary relationships are much contested by scientists. They include many large species, such as **Allosaurus** and **Giganotosaurus**.

The carnosaurs first evolved around 190 million years ago (the Early Jurassic period) and became common and widespread during the Jurassic and Cretaceous periods, before going extinct around 65 million years ago. The first dinosaur ever to be found (*Megalosaurus*, discovered in England and named in 1824) was a species of carnosaur. When the skeleton of *Eustreptospondylus* was first discovered it was mistaken for that of *Megalosaurus*, but in 1964 the mistake was corrected. The scarcity of its fossils suggests that, in its day, *Eustreptospondylus* would have been a rare sight.

Fascinating Fact > *Eustreptospondylus* probably lived on islands in warm seas, and may even have been a good swimmer.

Othnielia
A small, fast herbivore

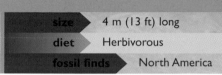

name		*Othnielia* (oth-NEE-eh-lee-ah), meaning 'belonging to Othniel'	size		4 m (13 ft) long
animal type		Ornithischian (ornithopod) dinosaur	diet		Herbivorous
lived		154–144 million years ago	fossil finds		North America

Othnielia was a small dinosaur that lived in woodland and other open areas during the Late Jurassic period. It probably lived in small herds and was entirely vegetarian. It had a horny beak that could crop low vegetation, and a battery of chisel-shaped teeth that could chew plant material before it was swallowed. *Othnielia* was bipedal, with small arms and a stiff tail for balance. This meant that it was fast and manoeuvrable, and it needed to be. Being small, and with few means of defence, *Othnielia* would have used its speed and cunning to escape from large predators, such as *Allosaurus*.

The first *Othnielia* fossils were discovered in 1877 in Colorado, and the animal was named after Othniel Marsh, one of the most celebrated palaeontologists of the nineteenth century. It belongs to a specialized group of dinosaurs known as hypsilophodontids,

whose many species tended to be quite small and nimble. Fossils of hypsilophodontid dinosaurs have been found in many parts of the world, and in some areas they are very common. The Isle of Wight, off the southern coast of England, has produced many well-preserved hypsilophodontid skeletons, believed to be from animals that became trapped in quicksand.

It used to be thought that some of the smaller hypsilophodontids lived in trees, grasping the branches with their long, clawed feet. This is now considered very unlikely, as they were far too heavy and awkward. At around 150 million years old, *Othnielia* is one of the oldest-known hypsilophodontid dinosaurs.

Top > *Othnielia* may have followed behind large herbivorous dinosaurs, such as *Stegosaurus*, feeding on the debris they left in their wake.

Diplodocus
The longest of the dinosaurs

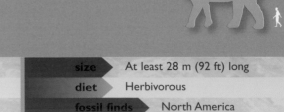

name	Diplodocus (dip-LOD-oh-kus), meaning 'two-beam reptile'	size	At least 28 m (92 ft) long
animal type	Sauropod (diplodocid) dinosaur	diet	Herbivorous
lived	154–144 million years ago	fossil finds	North America

Diplodocus is one of the most famous dinosaurs of all. It was a truly gigantic beast that lived on the drought-ridden plains of North America during the Late Jurassic period. The majority of its 28-metre (92-ft) length was made up of its long tail and neck, but even so, *Diplodocus* could reach a weight of 15 tonnes or more – about four times the weight of an elephant. To help support this vast bulk, *Diplodocus* had four upright legs built from enormous, column-like bones that directed its weight straight down to the ground.

It used to be thought that *Diplodocus* used its long neck to reach high into the trees to browse. However, it is now known that its neck was not very flexible, so could not have been raised very far. Rather than looking up for its food, *Diplodocus* probably spent its time sweeping the ground for foliage such as ferns, young trees and bushes. Its stout jaws and peg-like teeth were very efficient at denuding plants of their leaves and young shoots. The jaws, however, were not capable of chewing, so its stomach was full of smooth pebbles called gastroliths, which would break down tough plant matter. *Diplodocus* purposely swallowed small, smooth pebbles from riverbeds and lakesides.

Fossilized footprints suggest that *Diplodocus* lived in small herds that were constantly on the move, looking for fresh feeding grounds. During the mating season *Diplodocus* would scrape a semi-circular trench in soft ground and lay several large, round eggs inside it. They would then be covered up and left to hatch under ground. A newly hatched *Diplodocus* would have weighed about 7 kg (15.5 lb) and would probably have

Fascinating Fact > The end of *Diplodocus*'s tail was so long and thin that some scientists believe it was used like a bullwhip for defence.

fended for itself for a few years before joining an established herd.

The brain of *Diplodocus* was only fist-sized, but its body was so long that it had a secondary nerve centre in its lower back that was used to boost signals to the tail and hind legs. Even though it was enormous, *Diplodocus* still faced attack from large, predatory dinosaurs, such as **Allosaurus**. Its main means of defence were its whip-like tail and being part of a herd. However, old or injured *Diplodocus* that could not keep up with the herd soon fell victim to other dinosaurs.

The first *Diplodocus* bones were discovered in Colorado in 1877, with a complete skeleton being found some years later. Recent fossil finds suggest that it had spines running down its back. Some scientists think that *Diplodocus* and *Seismosaurus*, an even larger sauropod, were the same animal. If so, the largest *Diplodocus* would have been 45 m (147.5 ft) long and around 30 tonnes in weight.

Diplodocus was a sauropod, a group of vegetarian dinosaurs that are related to (but probably not descended from) the prosauropod dinosaurs, such as **Plateosaurus**. Together the sauropod and prosauropod dinosaurs form the sauropodomorphs, which, with the theropods and ornithischians, form the three main types of dinosaur.

Within the sauropods, there are many subgroups, including the diplodocids, to which *Diplodocus* belongs, the brachiosaurids (e.g. **Brachiosaurus**) and the titanosaurids (e.g. **Argentinosaurus**). The earliest sauropod species date from around 190 million years ago, and they flourished from then until the end of the Cretaceous period, 65 million years ago. The diplodocids were at their most diverse during the Late Jurassic and Early Cretaceous (159–99 million years ago), when more than 20 species are known from Europe, Africa and North and South America. After this time their numbers dwindled, probably because they were outcompeted by their close cousins the titanosaurids.

Left > A herd of large sauropods such as *Diplodocus* would have denuded an area of vegetation in a very short space of time.

Brachiosaurus
A treetop-feeding dinosaur

name	Brachiosaurus (BRAK-ee-oh-SAW-rus), meaning 'arm reptile'	size	23 m (75.5 ft) long
animal type	Sauropod (brachiosaurid) dinosaur	diet	Herbivorous
lived	155–112 million years ago	fossil finds	Africa and North America

Brachiosaurus was the heavyweight dinosaur of the Late Jurassic period. Although it was shorter than **Diplodocus** (another large dinosaur), *Brachiosaurus* was over five times heavier, weighing in at around 80 tonnes. To help it cope with this weight, *Brachiosaurus* had massive leg bones and vertebrae, and must also have had a strong heart in order to pump blood up its long neck to its head.

The front legs of *Brachiosaurus* were among the longest of any animal, being over 4 m (13 ft) in length; they were far taller than the rear legs, giving *Brachiosaurus* a steeply sloping back. This, combined with a long, upright neck, gave *Brachiosaurus* a head height of over 12 m (40 ft), allowing it to reach right up to the tops of the trees. Its skull had a line of sharp, chisel-like teeth that could crop away at the leaves and cones of conifer trees.

The first fragments of *Brachiosaurus* were discovered in 1900 in Colorado. More bones were found in the following years, and in 1908 further fossils were discovered in Tanzania. It used to be thought that *Brachiosaurus* lived under water, walking on the lakebed and breathing through its high nostrils, rather like a snorkel. This supposition arose because scientists felt that the animal needed the water to support its enormous weight. However, it is now known that this would have been physically impossible because the pressure of water on its lungs would not have allowed it to breathe.

Brachiosaurus was a brachiosaurid sauropod, which makes it a close cousin of diplodocids, such as **Diplodocus**, and titanosaurids such as **Argentinosaurus**. The brachiosaurids characteristically had long, upright necks, which allowed them to eat food that was beyond the reach of the lower-necked sauropods, such as **Diplodocus**. They first evolved around 170 million years ago; they were geographically widespread and went extinct around 95 million years ago, in the Late Cretaceous period.

Fascinating Fact > An adult *Brachiosaurus* could weigh the same as about 20 elephants.

Anurognathus
A tiny insectivorous pterosaur

name	Anurognathus (an-YOOR-og-NATH-us), meaning 'without tail and jaw'		size	50-cm (20-in) wingspan
animal type	Pterosaur reptile		diet	Insectivorous
lived	156–144 million years ago		fossil finds	Europe

Anurognathus was a minute pterosaur, whose body was only around 5 cm (2 in) long. In total, *Anurognathus* may have weighed less than a garden sparrow, but it still managed to have a wingspan of 50 cm (20 in) – about 10 times its body length. It lived by hunting small flying insects on the wing, catching them with its stout beak and needle-like teeth.

It has been speculated that *Anurognathus* might have lived near large sauropod dinosaurs, such as **Diplodocus**, whose messy eating habits and voluminous dung would have attracted many flying insects. They may even have perched on the back of these great dinosaurs, picking off any insects that lived on or around their skin.

Only one fossil specimen of *Anurognathus* is known. It was found in 1923 in one of the Solnhofen lime-stone quarries in Bavaria, which are famous for pro-

ducing specimens of extraordinary quality. That said, the single *Anurognathus* fossil, although complete, is not as perfectly preserved as some other Solnhofen specimens, such as **Rhamphorhynchus**.

Anurognathus is a bit of a puzzle to scientists. Although its skeleton reveals that it was closely related to pterosaurs, such as **Rhamphorhynchus** and **Peteinosaurus**, it seems to have lost the long tail that is so characteristic of these pterosaurs. In this respect it resembles (but is not the ancestor to) some of the later tailless pterosaurs, such as **Tapejara** and **Ornithocheirus**.

Fascinating Fact > Some insects may have been too large for this tiny pterosaur to fit in its mouth.

95

Stegosaurus
A famous spiky-tailed herbivore

name	Stegosaurus (STEG-oh-SAW-rus), meaning 'covered reptile'
animal type	Ornithischian (stegosaurid) dinosaur
lived	154–144 million years ago

size	9 m (30 ft) long
diet	Herbivorous
fossil finds	North America

The spiky-backed *Stegosaurus* is a very distinctive and famous dinosaur. It could reach the length of a bus and weighed around 7 tonnes. Fossilized footprints suggest it lived a solitary existence, wandering the woodlands and open plains in search of fresh leaves and shoots.

Stegosaurus was very slow-moving, so had to be able to protect itself from attack. Aside from its bulk and back plates, it also had a club tail containing four solid spikes each up to 1.2 m (4 ft) long, which it could swing at enemies. It also had armour-plated skin under its throat. Only the largest carnivorous dinosaurs, such as **Allosaurus**, would risk attacking a fully grown *Stegosaurus*.

This animal is famous for having a brain that was the size of a walnut and weighed just 70 g (2.5 oz). However, *Stegosaurus* also had a secondary nerve centre (sometimes incorrectly called a 'second brain') near the base of its spine that helped it move its rear legs and tail.

The two rows of diamond-shaped plates along *Stegosaurus*'s back have been the subject of much scientific speculation. It was thought that they were used to stop other dinosaurs from jumping on its back. However, it is now known that each plate was filled with hundreds of tiny blood vessels. If *Stegosaurus* needed to cool down, it could have pumped blood into the plates, which would have caused them to turn bright red. The red plates could also have been used as part of a mating display.

The first *Stegosaurus* fossils were discovered in Colorado in 1877, and its unusual shape and spiky plates ensured that it quickly became an iconic dinosaur. *Stegosaurus* is a stegosaurid and part of the ornithischian (bird-hipped) dinosaurs, a group that includes heavily armoured giants such as **Polacanthus** and **Torosaurus**.

Fascinating Fact > In relation to its size, *Stegosaurus* had the smallest brain of any dinosaur.

Although *Stegosaurus* is found only in North America, there were many related species of stegosaurid found in Africa, Europe and Asia. Of these, it is the stegosaurid fossils from China that have attracted particular interest, as these seem to have evolved in isolation from the other species in the world, producing new and interesting-looking species.

The earliest stegosaurid species were found in Europe and date from around 180 million years ago. The group was at its most abundant and diverse during the Late Jurassic and Early Cretaceous periods (159–132 million years ago), but after this time their numbers dwindled until, by 99 million years ago, they had become extinct. It used to be thought that the stegosaurids had survived through to the Late Cretaceous period, but the fossil on which this supposition was based, *Dravidosaurus*, turned out to be from a plesiosaur.

Above > Fossils of *Stegosaurus* surprised Victorian scientists and delighted the public. The dinosaur was even given a central role in Sir Arthur Conan Doyle's *The Lost World* (1912).
Top left > *Stegosaurus*'s spiked tail may have been heavy, but as a means of defence it would have been unbeatable.

Allosaurus
The lion of the Jurassic

name	Allosaurus (all-oh-SAW-rus), meaning 'different lizard'		size	12 m (40 ft) long
animal type	Theropod (carnosaur) dinosaur		diet	Carnivorous
lived	154–144 million years ago		fossil finds	North America, Europe and possibly Africa

The largest predator in the Late Jurassic period was the gigantic *Allosaurus*. It was a bipedal hunter armed with powerful jaws, long teeth and sharp claws. When hunting alone, it would have attacked small to medium-sized dinosaurs, but when pack-hunting, several *Allosaurus* would have been capable of bringing down very large dinosaurs, such as **Diplodocus**.

During an attack, *Allosaurus* would have used its short but strong arms to hook on to the body of its prey, and then draw it in towards its mouth. *Allosaurus* had a jaw lined with sharp, serrated teeth, some of which were 10 cm (4 in) long. A series of deep bites would soon cause any prey to die from loss of blood, after which *Allosaurus* could feed at will.

Unlike some of the later theropod dinosaurs (including **Tyrannosaurus**), *Allosaurus* did not have jaws that were powerful enough to crunch through bone. Instead, it was forced to eat only soft parts of the body, such as the stomach and throat. This meant that after it had finished eating there would be plenty of meat left on the carcass for pterosaurs, smaller dinosaurs and other scavenging animals to pick at.

Female *Allosaurus* were larger than the males, and, if they followed modern reptile behaviour, were probably extremely aggressive even to members of their own species. During the mating season, males would

Below > *Allosaurus* may have cooperated with one another in order to bring down large dinosaurs such as these *Diplodocus*.

Fascinating Fact > *Allosaurus* was the top predator in Sir Arthur Conan Doyle's novel *The Lost World* (1912).

have to approach the females with caution, and leave soon after mating had taken place. It is thought that the female would have buried her eggs in a mound of soil and vegetation, and young *Allosaurus* would have grown quickly, reaching maturity in only a few years.

The first, rather scrappy, *Allosaurus* fossils were found in 1877 in Fremont County, Colorado, and it was several years before more complete skeletons were recovered. In 1927 a quarry in Utah was found to contain the skeletons of over 40 *Allosaurus* of all ages. It is thought that they had drowned in a muddy swamp while trying to get at the bodies of other mired animals. Close examinations of *Allosaurus* skeletons show that they led a very rough life indeed, and were prone to injury and infection. One particularly complete *Allosaurus* skeleton (nicknamed 'Big Al') reveals that during its life the unfortunate animal picked up dozens of injuries, especially to its feet. Eventually, these injuries would have left this dinosaur badly crippled and unable to fend for itself: it probably died of starvation or thirst.

Allosaurus belongs to the carnosaurs, a successful group of large, predatory therapod dinosaurs, which included the largest land carnivore of all time, **Giganotosaurus**, and the coastal-dwelling **Eustreptospondylus** (see the latter for more details about the carnosaurs).

The theropods were arguably the most successful of all the dinosaur groups. For the most part, they were bipedal, swift-moving carnivores, but they came in many different shapes and sizes, including bizarre forms, such as the herbivorous **Therizinosaurus** and the bird-like **Mononykus**. The theropods are among the oldest known of the dinosaurs (see **Coelophysis**) and prospered throughout the Triassic, Jurassic and Cretaceous periods, producing everything from the chicken-sized *Compsognathus* to the famously large **Tyrannosaurus**. The theropods also gave rise to the birds (see **Ornitholestes**), a group that managed to survive the great extinction 65 million years ago. In this respect, the theropods were the only dinosaur group to leave descendants that can still be seen on the Earth today.

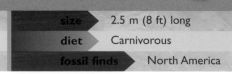

name	Ornitholestes (or-NITH-oh-LESS-teez), meaning 'bird thief'	size	2.5 m (8 ft) long
animal type	Theropod (coelurosaur) dinosaur	diet	Carnivorous
lived	154–144 million years ago	fossil finds	North America

Ornitholestes was a small, bipedal dinosaur that hunted small reptiles and insects. It was light and capable of moving with speed and agility through wooded areas and across open plains. Its long, stiffened tail accounted for nearly half its body length, and was used as a counterbalance when running or jumping. Its small size would have made it vulnerable to attack from large predators, such as ***Allosaurus***.

As a hunter of small animals, *Ornitholestes* probably lived a solitary life and may have met with others of its kind only in order to mate. Its small nose horn and head crest were probably used as part of a courtship display.

Only two partially complete fossils of *Ornitholestes* have been found, both from the same quarry in Wyoming. *Ornitholestes* is a coelurosaur, part of a large group of carnivorous dinosaurs that includes ***Coelophysis***. The coelurosaurs are the oldest group of theropod dinosaurs, with most being small, light-framed and agile (see ***Coelophysis***).

The coelurosaur dinosaurs have been of great interest to scientists because their skeletons closely resemble those of birds. For many decades the fossil origins of birds was a controversial topic. It used to

be thought that they had evolved within the last 65 million years; then, in 1861, a beautifully preserved fossil of a bird was discovered in Bavaria. The fossil was 145 million years old (Late Jurassic period) and was named *Archaeopteryx* (meaning 'ancient wing'). This fossil showed an animal that was not an ordinary bird of the sort we see today: it had many reptilian features, such as a long tail, claws on its wings and rows of sharp teeth in its beak. Were it not for the fact that *Archaeopteryx* had feathers, it would have been misidentified as a type of dinosaur.

There was an obvious close resemblance between the skeleton of *Archaeopteryx* and that of the therapod dinosaurs. This led scientists to speculate that birds had evolved from the dinosaurs. However, finding proof of a connection between birds and dinosaurs proved to be problematic, and several alternative theories were suggested. The issue began to be resolved in the 1970s, when studies of several coelurosaur dinosaur species (such as *Deinonychus* and ***Velociraptor***) revealed similarities between this group of dinosaurs and birds.

The matter of the birds' origins was all but settled with the discovery of the so-called 'feathered dinosaurs' in the Liaoning region of China. In the 1990s the Liaoning quarries produced a series of near-perfectly preserved fossils of some small but advanced species of coelurosaur dinosaur. This includes species such as *Sinosauropteryx* and *Protoarchaeopteryx*.

In life, these new dinosaurs all appear to have had a thin covering of feathers (or something like feathers), which they probably used to keep them warm or for display purposes. One small, crow-sized species, *Microraptor*, may even have been able to climb trees and to glide down to earth. The occurrence of feathers on some coelurosaur dinosaurs proves to most scientists' satisfaction that birds did evolve from dinosaurs.

Ornitholestes lived at a time when birds were first evolving and it had many bird-like characteristics. Its bones were lightweight and hollow, and it could fold its fearsome claws and arms into its body (like wings). Some reconstructions even give *Ornitholestes* feathers, which it may have used for warmth or for making colourful mating displays. (For other aspects of the birds' relationship with dinosaurs see ***Velociraptor*** and ***Mononykus***.)

Left > The reptilian features of *Archaeopteryx* first alerted scientists to the idea that birds and dinosaurs are closely related.

Iguanodon
A herbivore with a thumb spike

name	*Iguanodon* (ig-WAN-oh-DON), meaning 'iguana tooth'		size	10 m (33 ft) long
animal type	Ornithischian (ornithopod) dinosaur		diet	Herbivorous
lived	144–112 million years ago		fossil finds	Europe, Asia and North America

Iguanodon was one of the most successful dinosaurs of its day. It had a large, heavy body, a sharp beak for cropping vegetation, and rows of flattened teeth for chewing. It may have lived in sizeable herds, and could survive in a variety of habitats, from semi-desert to lush forests.

Fossilized footprints show that *Iguanodon* walked on all fours, but it probably reared on to its back legs when feeding to help it reach higher into the trees. This also freed its hands, which could grasp vegetation and drag it closer towards the mouth. It is thought that an adult *Iguanodon* could have eaten up to 135 kg (300 lb) of vegetation a day.

Having a hinged jaw allowed *Iguanodon* to grind vegetation between its teeth, and being able to chew greatly speeded up the body's ability to digest food. This gave *Iguanodon* a big advantage over the other herbivores around it.

Other than its size, *Iguanodon* had few means of defence. It had a thumb spike that could have been used as a weapon at close quarters. To do so it would have needed to rear up on two legs, which alone would have scared off many potential attackers.

Fossils of *Iguanodon* were first found in southern England in 1822. It was one of the first dinosaurs to be identified by scientists, but early attempts at reconstructing its skeleton were a disaster and made it look like a rhino (the thumb spike was mistakenly placed on the nose). It was not until 1878, and the chance discovery of over 30 near-complete skeletons in Belgium, that *Iguanodon* was reconstructed with an upright stance.

Iguanodon was just one of several species of iguanodontid that lived between 160 and 65 million years ago. The iguanodontids were a group of large and herbivorous dinosaurs, whose fossils are known from across the globe (with the exception of Antarctica). They were ornithopod dinosaurs, a subgroup of dinosaurs within the ornithischia that includes the hypsilophodontids (see **Othnielia**) and the hadrosaurs (see **Anatotitan**).

The ornithischia are the third great group of dinosaurs, the other two being the theropods and sauropodomorphs. The evolutionary relationship between the ornithischia, theropods and sauropodomorphs is unclear, but their common ancestor probably lived in the Middle Triassic period (242–227 million years ago). The earliest ornithischian dinosaurs are the heterodontosaurids, a group of small, lightly armoured herbivorous dinosaurs that lived around 190 million years ago. All the ornithopod dinosaurs, such as *Iguanodon*, **Othnielia** and **Anatotitan**, are descended from the heterodontosaurids.

Fascinating Fact > *Iguanodon* was the second dinosaur to be identified by science (the first was *Megalosaurus*).

Right > *Iguanodon's* sharp beak and rows of flat teeth could crop plants and chew them into a pulp, making them easier to digest. Its long neck helped it reach high into the trees.

Below > The fossilized footprints of *Iguanodon* are commonly found in many parts of the world and suggest that these animals travelled in herds. This would have given them some protection against large predators.

The ornithischians were probably the most diverse of the dinosaurs, producing hundreds of species that included a variety of different body shapes, defence mechanisms and lifestyles. In addition to large ornithopods, such as *Iguanodon*, the other main groups of ornithischian dinosaurs are the horned ceratopia (e.g. **Torosaurus**), the spiny stegosauria (e.g. **Stegosaurus**), the heavily armoured ankylosauria (e.g. **Polacanthus**) and the hard-headed pachycephalosauria (e.g. *Pachycephalosaurus*).

Tapejara
A bizarre crested pterosaur

name	Tapejara (TAP-eh-JAR-ah), meaning 'the old being'		size	5-metre (16.5-ft) wingspan
animal type	Pterosaur reptile		diet	Carnivorous
lived	121–112 million years ago		fossil finds	South America

Tapejara was a medium-sized pterosaur that hunted fish along the ancient Cretaceous coastlines of South America. It is best known for its large head crest, which could be up to 1 m (3.3 ft) in height. The crest, which may have been found only on males, would have been a great disadvantage to the animal, making its head heavy and its flight very slow and unstable. Instead, the crest would almost certainly have been used as part of an elaborate courtship display, and would therefore have been brightly coloured to attract the attention of females. *Tapejara* nested in large colonies near to the coast, where the parents could catch fish to feed to their young.

Above > Fossils from Brazil show that the crest of *Tapejara* could be up to three times as tall as its skull.

So far, *Tapejara* is the oldest-known pterosaur to have possessed a beak without any teeth. It, and one other similar toothless species, is known from the Santana Formation of Brazil, a suite of rocks that has produced many fantastic pterosaur fossils. Without further fossils, it is difficult to say exactly how *Tapejara* is related to the many species of toothless pterosaur (such as **Pteranodon**) that lived during the Late Cretaceous period.

Fascinating Fact > *Tapejara*'s head crest was so large that it would have been highly vulnerable to crosswinds.

Polacanthus
A heavily armoured herbivore

name	Polacanthus (POLE-ah-KANTH-us), meaning 'multiple spines'	size	4 m (13 ft) long
animal type	Ornithischian (ankylosaurid) dinosaur	diet	Herbivorous
lived	137–121 million years ago	fossil finds	Europe

Polacanthus was a low-slung, heavy-set and well-armoured dinosaur that browsed on low trees and bushes. It was four-legged and slow-moving, which would have made it a potential target for large predators. For this reason, its back and shoulders sported rows of stout spikes that were angled outwards to provide maximum protection. Its tail was also heavily armoured, and could have been swung backwards and forwards to ward off potential attackers. *Polacanthus* was particularly heavily protected over its hips, as these would have been the obvious target for a large carnivore.

It is thought that *Polacanthus* was a solitary animal, wandering randomly in search of leaves and other soft foliage. Some have suggested that it may have followed behind other large, plant-eating dinosaurs, such as *Iguanodon*, feeding off the vegetation they exposed or left behind.

Although first discovered in 1865, fossils of *Polacanthus* are known only from southern England, and a complete skeleton has yet to be found. It was part of the ankylosauria, a subgroup of heavily armoured dinosaurs within the ornithischian dinosaurs (see *Iguanodon* for more details). As such, *Polacanthus* is distantly related to other large, well-armoured herbivores, such as *Stegosaurus* and *Torosaurus*.

The ankylosaurid dinosaurs first evolved around 180 million years ago (the Middle Jurassic period), but did not become more numerous until the end of the Early Cretaceous period (100 million years ago). The ankylosaurids were never very numerous, but they did produce some of the most heavily armoured dinosaurs of all (e.g. *Ankylosaurus*). As a group, they survived until the mass extinction event that killed all the dinosaurs, 65 million years ago.

Fascinating Fact > Although *Polacanthus* was a small dinosaur, standing only 1 m (3.3 ft) tall, it weighed over a tonne because of its bony armour.

Iberomesornis
An early bird

name	*Iberomesornis* (eye-BER-oh-mes-OR-nis)	size	10–15-cm (4–6-in) wingspan
animal type	Iberomesornithid bird	diet	Carnivorous
lived	137–121 million years ago	fossil finds	Europe

As one of the oldest-known true birds, *Iberomesornis* had feathers, wings and a shortened tail, just like many modern birds. However, it also had some features not seen in modern birds, such as claws on its wings and a primitive pelvis.

Iberomesornis was only the size of a large sparrow, but it was obviously an accomplished flier, and probably hunted insects and small animals, plucking them out of the air or from the ground. It may have preferred to live near to lakes, catching insects from the water's surface. Despite being able to perform many aerial feats, such as turning and swooping at speed, it probably couldn't manoeuvre at slow speeds. When resting, it used its long, clawed feet to perch on branches. It is unlikely to have been able to sing, although it could probably have made basic squawking noises.

The first (and only known) fossil of *Iberomesornis* was discovered in 1985 in Las Hoyas, Spain. Although other fossilized birds are known from this locality, *Iberomesornis* is particularly important to scientists because it belongs to the Enantiornithine (or 'opposite') birds. For many years there was a massive gap in the fossil record of birds between the Early Jurassic species *Archaeopteryx* (see **Ornitholestes** for details) and the 'toothed birds' of the Late Cretaceous period (see **Hesperornis**). The Enantiornithine birds were known to bridge this gap in the fossil record but prior to the discovery of *Iberomesornis* little was known about them.

Most other birds from this time (around 135 million years ago) have reptilian characteristics, such as a long tail. *Iberomesornis* does not, and its skeleton shows that even in the Early Cretaceous period, some birds were beginning to move away from their dinosaur ancestry towards the modern bird shape that we are used to.

Since the discovery of *Iberomesornis*, many more Enantiornithine fossils have come to light in the quarries of Liaoning province, China. In only a few years the number of known fossilized Enantiornithine bird species has grown exponentially. Well-preserved specimens of Chinese species such as *Sinornis* and *Confuciusornis* have permitted scientists to understand the birds' early fossil history better. The Enantiornithine birds went extinct around 65 million years ago, after which only the Neornithine (or modern) birds remained.

Utahraptor
A sickle-clawed killer

name	Utahraptor (YOO-tah-RAP-tor), meaning 'Utah thief'		size	6 m (19.5 ft) long
animal type	Theropod (coelurosaur) dinosaur		diet	Carnivorous
lived	127–121 million years ago		fossil finds	North America

Utahraptor was a swift and terrifying predatory dinosaur that stood taller than a man. It had a number of features that made it an especially efficient and deadly killer. Its skeleton was lightly built, making it swift and manoeuvrable, while its long, stiffened tail helped it to balance. This would have allowed it to perform acrobatic feats, such as jumping and balancing on one foot.

The most deadly feature of *Utahraptor* was its long clawed hands and feet. The second foot claw was 20 cm (8 in) long, razor sharp and could be rotated through a wide arc. When attacking a large animal, such as another dinosaur, *Utahraptor* would have leapt at it feet first, dragging the enlarged second claw down its flank and ripping at the flesh as it did so. The wounded prey would then have been systematically killed by *Utahraptor*, which would have used its arms to hold it still or to rake at it further.

As its name suggests, *Utahraptor* was first discovered in Utah where it is known from a single skeleton. It was a coelurosaur dinosaur, which places it in the same group as other small and deadly theropod dinosaurs, such as *Coelophysis* and *Ornitholestes*. However, within the coelurosaurs there was another subgroup of dinosaurs, known as the dromaeosaurids. These are characterized by their light and agile skeletons, keen senses, and by the deadly sickle-like claw on each foot.

Utahraptor was the largest of several species of dromaeosaurid dinosaur, all of which were deadly predators (see *Velociraptor*). The discovery of several dromaeosaurid dinosaurs surrounding a single large, plant-eating dinosaur suggests that they may have hunted in packs, repeatedly attacking a single dinosaur until it died from blood loss.

Fascinating Fact > *Utahraptor* was discovered while the film *Jurassic Park* was being shot; it helped to justify the excessively large size of the film's *Velociraptor* dinosaurs.

Ornithocheirus
An aircraft-sized pterosaur

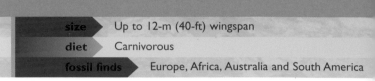

name	Ornithocheirus (AWN-ith-oh-KY-rus), meaning 'bird-like hand'	size	Up to 12-m (40-ft) wingspan
animal type	Pterosaur reptile	diet	Carnivorous
lived	140–70 million years ago	fossil finds	Europe, Africa, Australia and South America

Ornithocheirus is possibly the largest known of all the pterosaurs. It had a long, slender skull and a long, tapering jaw lined with sharp teeth. It appears to have lived in coastal regions or around large inland water bodies, such as lakes and floodplains.

Using rising air currents, the largest *Ornithocheirus* specimens may have been able to fly hundreds of kilometres without once flapping their wings. As fish-eaters, they would have swooped down from on high, skimming across the water's surface. When a fish was spotted, the beak would be momentarily dipped under the surface, the jaws would close and the fish swallowed whole. The large, keel-like crest on the head may have helped the beak keep a straight path when dipped in the water.

Below > The Cretaceous skies were home to the largest flying animals of all time. With their low body weight and wide, skin-covered wings, the pterosaurs were ideally suited to gigantism.

Like most pterosaurs, *Ornithocheirus* would have been slow and awkward when moving on land, making it vulnerable to attack. It probably nested in large colonies close to the seashore. This would have provided safety in numbers, but may have meant that prime nesting sites, such as clifftops and beaches, were extremely crowded.

The first *Ornithocheirus* bones were discovered in 1827 in southern England, but it was not named until 1869. There are about 30 known species, but it is suspected that many of these may have been misidentifications of other pterosaurs. Therefore, although *Ornithocheirus* specimens have been reported from all the world's continents (apart from North America and Antarctica), it may turn out that some of these records are erroneous. Most species of *Ornithocheirus* had a wingspan of 2.5 m (8 ft) or less, but the bones of what appears to be a gigantic species were recently found in the Santana Formation of Brazil.

Fossils of pterosaurs have been known to scientists over 250 years but *Ornithocheirus* was one of the first to be given a formal scientific name. In the mid-eighteenth century complete skeletons of small pterosaurs were recovered from the Jurassic Solnhofen limestone of Bavaria, but these animals looked so unlike anything alive today that for many years nobody really knew what to make of them. Some people thought they were a type of duck, others that they were miniature dragons or bats. Only in 1809 did

Fascinating Fact > *Ornithocheirus* was the size of a small aeroplane, yet because of its hollow bones its body probably weighed less than a human being.

the great French anatomist Georges Cuvier finally work out that they were a form of extinct flying reptile that he called a pterodactyl ('wing finger').

Later in the nineteenth century a battle emerged between two leading British scientists, Richard Owen and Harry Seeley, over whether the pterosaurs were cold-blooded reptiles (as Owen believed) or warm-blooded 'primitive birds'. The discovery of many more pterosaur fossils (some of them fantastically well preserved) helped settle the issue. The pterosaurs were certainly reptiles (as Owen believed), and not related to birds, but there is still a debate as to whether or not they were warm-blooded (see *Rhamphorhynchus*). In fact, pterosaurs belong to the archosaurs, a broad group of reptiles that also includes the dinosaurs and crocodiles (see *Peteinosaurus*).

Koolasuchus
A relic from the past

name	*Koolasuchus* (KOOL-ah-SOOK-us), meaning 'Kool's crocodile'
animal type	Temnospondyl amphibian
lived	137–112 million years ago

size	5 m (16.5 ft) long
diet	Carnivorous
fossil finds	Australia

Koolasuchus was a gigantic amphibian with a large, flat and spade-shaped head. It lived in deep rivers, on floodplains and in lakes around the South Pole, where it hunted fish, crabs, turtles and medium-sized reptiles. Its head was 50 cm (20 in) long, and was large in comparison to its thin body and tail. The jaws were stout and lined with over 100 teeth, some 10 cm (4 in) long. The eyes, being on top of the head, gave *Koolasuchus* good all-round vision, and its skull had a series of grooves running across it that were filled with nerve tissue, allowing it to sense vibrations in water. All this enabled it to hide in muddy water, waiting for fish or crustaceans to pass within striking range.

When under water, *Koolasuchus* was very manoeuvrable, but on land its thin legs would have made moving very difficult, and it was thus vulnerable to predators. Its slimy skin and fish-like gills ensured that it could never have strayed far from water.

The first *Koolasuchus* fossils were discovered in 1989 in the south Australian fossil site known as Dinosaur Cove. During the Early Cretaceous,

Dinosaur Cove was in a subpolar region, and may have been frozen over for part of the year. *Koolasuchus* would probably have hibernated under the ice during these times.

Koolasuchus belongs to the temnospondyls, a group of amphibians that includes ***Proterogyrinus*** and *Rhinesuchus*. The temnospondyl amphibians once dominated the swamp forests of the Carboniferous period (354–290 million years ago), but the drier conditions of the Permian period did not suit them, and they went into a steep decline.

It was thought that all the temnospondyls had gone extinct in the Late Triassic, around 230 million years ago, but the discovery of *Koolasuchus* has extended their time range by around 100 million years to the

Fascinating Fact > *Koolasuchus* had a mouth that acted like a fish slicer; it even had long teeth in the top of its palate.

Early Cretaceous period. Many scientists believe *Koolasuchus* is an example of a 'relic fauna' that somehow managed to survive in the extreme environment of the South Pole long after all the other temnospondyls were driven to extinction elsewhere.

Even allowing for *Koolasuchus*, the extinction of the last temnospondyl probably occurred around 100 million years ago. After this the only amphibians left on Earth were species belonging to the lissamphibia or modern amphibians. Lissamphibians include all the amphibians alive on Earth today, such as frogs, newts, salamanders and the worm-like caecilians. Because they are small and delicate animals, their fossils are very rare; the oldest-known lissamphibian fossil – a frog – dates from the Late Triassic (around 215 million years ago) of Madagascar. The origins of the lissamphibia are obscure, but it is suspected that they evolved from a small and unusual group of temnospondyl amphibians known as the branchiosaurs, which lived during the Permian period. The largest living amphibian is the Japanese giant salamander, which can reach lengths of 1.8 m (6 ft).

Above > *Koolasuchus* was a giant among amphibians with a massive head full of teeth and palatial tusks. It lived in the southern polar region, which in the time of the dinosaurs was thickly forested and had no permanent snow.
Below left > *Koolasuchus* hunted rather like a modern crocodile.
Below > *Rhinesuchus* was a large amphibian that lived in South Africa during the Late Permian, 250 million years ago. It was thought to be capable of burying itself in mud and hibernating in times of drought. *Rhinesuchus* was believed to be one of the last of the temnospondyl amphibians until the discovery of the Cretaceous *Koolasuchus* extended their known time range by 100 million years.

Leaellynasaura
A mysterious polar dinosaur

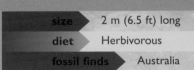

name	Leaellynasaura (Lee-EL-in-a-SAW-rah), meaning 'Leaellyn's reptile'		size	2 m (6.5 ft) long
animal type	Ornithischian (ornithopod) dinosaur		diet	Herbivorous
lived	112–99 million years ago		fossil finds	Australia

Leaellynasaura was a small but extraordinary dinosaur that lived in dense polar forests near the South Pole during the Early Cretaceous period. It moved on two legs, but may have dropped on to all fours when browsing on low plants, such as ferns and cycads.

The most amazing aspect of the *Leaellynasaura* lifestyle was the fact that it lived inside the Antarctic Circle, and was too small to migrate north during the winter. This meant that it would have had to survive several months of continual darkness and subzero temperatures – conditions that would not favour a small, cold-blooded dinosaur.

Quite how it did survive is not known; its large eyes may have helped it see in low light, and it also had small bumps at the back of its skull that could have accommodated enlarged optic lobes. It has also been suggested that *Leaellynasaura* may have hibernated during the coldest winter months when there would have been snow on the ground and few plants to eat. During the spring and summer months *Leaellynasaura* would have used the near-24-hour daylight to feed constantly, taking advantage of the warmer weather to grow and put on weight. The advantage of living in such an extreme climate was that there were few competitors for food and almost no predators around.

Like modern polar animals, *Leaellynasaura* would have had only a few weeks in the late spring and summer to mate, lay eggs and (presumably) raise its young. It would have been a struggle for both the newly hatched dinosaurs and the adults to eat enough food to allow them to survive during the cold winter months. The mortality rate among those juvenile dinosaurs experiencing their first winter must have been extremely high.

Fragmentary fossils of *Leaellynasaura* have been found at Dinosaur Cove in south Australia. From these it is possible to see that it is part of the hypsilophodontid group of ornithopod dinosaurs, which makes it a distant relative to **Othnielia**. The fossils are abundant, and they are just one of several similar species of hypsilophodontid to be found at Dinosuar Cove. This is unusual, as in other parts of the world hypsilophodontid fossils are quite rare; it suggests that the hypsilophodontids managed to carve a niche for themselves in the harsh Antarctic environment, an area where few other dinosaurs could venture. Fossils also show that some *Leallynasaura* managed to live for some time with infected leg wounds. This means that there must have been very few predators about, as any weak or injured animal would normally have been quickly picked off.

Above > This view of the top of a *Leaellynasaura* skull shows that this animal had very large eye orbits, suggesting that it could see well in the dark. This makes sense as *Leaellynasaura* would have spent much of its year in the low light and darkness of a polar winter.

Left > The polar summer would have given *Leaellynasaura* a matter of weeks in which to breed and raise its young. The most efficient way to do this would have been for the adults to care for and feed their newly hatched offspring.

In addition to the fossils from south Australia, other polar dinosaur fossils have been recovered from the Colville River region of Alaska, which, 90 million years ago, was well inside the Arctic Circle. The discovery of dinosaurs from the North and South Pole areas has interested scientists, as it brings into question some aspects of the asteroid impact theory that allegedly killed the dinosaurs 65 million years ago. If some dinosaurs could survive several months of cold and darkness, why didn't at least a few species survive the so-called 'nuclear winter' that followed the asteroid impact? Leaving such questions aside, the discovery of the polar dinosaurs shows just how successful and versatile these large reptiles really were.

Fascinating Fact > *Leaellynasaura* is named after Leaellyn, the daughter of Tom Rich, its discoverer.

Sarcosuchus
A giant crocodilian that ate dinosaurs

name	Sarcosuchus (SAR-koh-SOO-kuss), meaning 'flesh crocodile'		size	12 m (40 ft) long
animal type	Crocodilian reptile		diet	Carnivorous
lived	121–93 million years ago		fossil finds	Africa and South America

Sarcosuchus was a gigantic reptile that was twice as long as any present-day crocodile and weighed in at a massive 8–10 tonnes. Its skull alone was 1.6 m (5 ft) long. It was a crocodilian, not a true crocodile, but behaved much like many large crocodiles, living in rivers and large lakes, and coming ashore to lay eggs and rest in the sun.

As a meat-eater, *Sarcosuchus* was probably not fussy about its diet. It would have eaten a wide range of animals, from small and medium-sized dinosaurs through to fish, turtles, pterosaurs or anything else that strayed within its range. To catch land animals, *Sarcosuchus* would lie in ambush just beneath the water, with only its upward-facing eyes protruding. When an animal came to drink from the water's edge, *Sarcosuchus* would lunge forwards, grabbing it by the head, then pulling it into the water, where it would be drowned and then eaten. The mouth of *Sarcosuchus* contained over a hundred teeth, which, together with the hooked snout, made it difficult for any animal to escape from the jaws.

The unusual bulbous end on the snout provided *Sarcosuchus* with an excellent sense of smell, which may have been useful in detecting nearby animals. The nose also acted as a sound chamber, giving it a loud, deep call that may have been used during courtship displays. Fossils suggest that *Sarcosuchus* may have lived to quite an age – up to 60 years in some cases.

The first *Sarcosuchus* skull was found in 1964 in the Sahara desert, but it was not until the 1970s that a more complete skeleton was found. Since then its remains have also been discovered in Brazil, supporting the idea that South America and Africa were once joined together.

Sarcosuchus is part of the neosuchia (new crocodile) crocodilians, which were widespread and successful from the Early Jurassic through to the end of the Cretaceous period (around 206–65 million years ago).

Aside from giants such as *Sarcosuchus*, the neosuchians had many smaller species that lived in rivers and lakes, as well the sea.

Around 90 million years ago the first creatures belonging to the group eusuchia (true crocodiles) appeared, and from this evolved the three living crocodile families: crocodiles, alligators and gavials. During the Late Cretaceous period the eusuchians, in the form of the crocodiles and alligators (the gavials did not evolve until around 50 million years ago), became abundant and widespread, and could be found in most environments, including some Arctic regions. Among these eusuchian species was *Deinosuchus*, a Late Cretaceous crocodile that, in terms of size, rivals *Sarcosuchus*.

The mass extinction event at the end of the Cretaceous badly affected all the world's reptiles, including the crocodilians. Only the eusuchian crocodilians survived, but they prospered and are still with us today. The world's largest living reptile is the 7-m (22-ft) saltwater crocodile, which can weigh up to a tonne and has been responsible for many deaths and injuries in Australia. Just as well none of its ancestors are with us.

Left > The giant crocodilian *Sarcosuchus* makes a successful attack on an iguanodontid dinosaur that has come to the water's edge to drink. *Sarcosuchus* is almost twice as long as the biggest crocodiles alive today and would have been one of the few animals that could tackle large dinosaurs.

Right > *Sarcosuchus* was about the same size as the crocodilian *Deinosuchus*, which lived in the region of Texas during the Late Cretaceous period. Exactly which of the two was the bigger is a matter of debate.

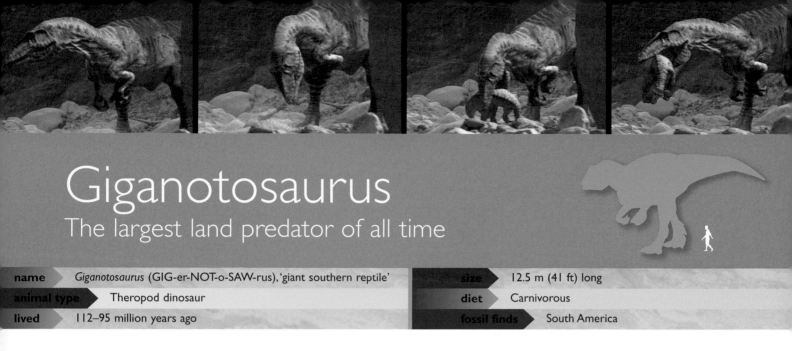

Giganotosaurus
The largest land predator of all time

name	Giganotosaurus (GIG-er-NOT-o-SAW-rus), 'giant southern reptile'	**size**	12.5 m (41 ft) long
animal type	Theropod dinosaur	**diet**	Carnivorous
lived	112–95 million years ago	**fossil finds**	South America

Giganotosaurus was the largest meat-eating dinosaur of all time. It was longer and stockier than **Tyrannosaurus**, and, at around 7 tonnes, much heavier. It was a formidable predator, with a long, narrow skull and scissor-shaped jaws that would have helped it catch and dissect prey with surgical precision. It fed on other large dinosaurs, and it is estimated that it would have needed to eat around 7 tonnes of meat a year to stay alive.

Like many predatory dinosaurs, *Giganotosaurus* had a keen sense of smell and good eyesight. It could quite possibly have smelt potential prey (or, indeed, a dead body) from some kilometres away. When close enough, it would put on a sudden burst of speed and launch an attack. *Giganotosaurus* may have been able to run at up to 50 kph (31 mph). However, quite how often it would have needed to use its running ability is a matter of debate, as much of its prey consisted of slow-moving sauropods, such as juvenile **Argentinosaurus**, or the 13-m (42.5-ft) long *Andesaurus*.

Recently, six *Giganotosaurus* fossils were discovered together in the same rock outcrop in Argentina. The animals, a mixture of adults and juveniles, were probably killed and then buried by a sudden flash flood that swept down a river valley, catching them by surprise. The fact that these animals were buried together suggests that *Giganotosaurus* travelled in small groups that contained both younger and older individuals. They may also have hunted in packs, picking on weak or injured animals, collectively killing them, and sharing the feast afterwards. Microscopic studies of their bones indicate that after hatching *Giganotosaurus* may have grown quickly, reaching adulthood in only 5–8 years.

It was only in 1993 that the first few *Giganotosaurus* bones were found in the Neuquen basin region of Argentina by an amateur fossil hunter. It was immediately apparent that the fossils belonged to a gigantic predatory dinosaur, and after further remains were discovered the animal was formally named in 1995. It is still known from only a few specimens, all discovered in the same region. The bones of *Giganotosaurus* have been chemically analysed in order to answer some questions about its biology and lifestyle. The results suggest that adults could have maintained a reasonably constant body temperature (sometimes called homeothermy or gigantothermy) by using their vast bulk to retain heat in their bodies, as opposed to generating heat, like mammals, or absorbing it from the sun, like most reptiles. Being able to retain heat in this way would have been advantageous to dinosaurs such as *Giganotosaurus*, as it would mean they didn't need to sunbathe in order to get warm enough to move, yet their energy requirements were still reasonably low (unlike mammals, whose energy requirements are very high). Using this information, it is possible to estimate that adult *Giganotosaurus* required around 20 kg (44 lb) of meat a day to live (this is about the equivalent of eating a medium-sized dog).

Prior to the discovery of *Giganotosaurus*, the record for the largest meat-eating dinosaur was held either by **Tyrannosaurus** or the African *Carcharodontosaurus*, depending on whose statistics you choose to believe. *Giganotosaurus*, however, is nearly a metre longer than either of these animals, and is much stockier and heavier as well. Before its discovery, few large predatory dinosaurs were known from South America.

Although there is still some uncertainty about its evolutionary history, *Giganotosaurus* is generally thought to be part of the carnosaur group of theropods (see **Eustreptospondylus** for details). Its skeleton shares many features with the large Jurassic dinosaur **Allosaurus**, which has led some people to suggest that *Giganotosaurus* is an allosaurid, a group of predatory dinosaurs noted for their large size and powerful jaws. The allosaurids were successful during the Late Jurassic and Early Cretaceous periods, after which their role was taken over by other large dinosaurs such as the tyrannosaurids (see **Tyrannosaurus**).

Top > *Giganotosaurus* was so large it would have been able to lift up and carry a medium-sized dinosaur in its jaws.

Argentinosaurus
The heaviest of all the dinosaurs

name	Argentinosaurus (AR-jen-TEE-noh-SAW-rus), 'Argentinian reptile'		size	35 m (115 ft) long
animal type	Sauropod (titanosaurid) dinosaur		diet	Herbivorous
lived	112–95 million years ago		fossil finds	South America

Even among the giant sauropod dinosaurs, *Argentinosaurus* was an exceptionally large animal. It had an overall body length of 35 m (115 ft) and weighed in at around 90 tonnes, making it the heaviest land animal ever to have lived. Like other sauropods, it was a vegetarian, and used its long neck to sweep the ground for ferns and bushes, or to reach into the conifer trees. Food had to travel such a long way through its huge body that the stomach and digestive system had plenty of time to ferment and break down the tough leaves and shoots. The stomach may also have contained smooth pebbles called gastroliths that would have helped pulverize the food.

Argentinosaurus probably moved in groups of 15–20 animals that included both juveniles and adults. The younger ones were vulnerable to attack from large theropod dinosaurs, such as *Giganotosaurus*, while the heavier adults would have had to watch their step when on swampy or unstable ground lest they slip or get stuck in soft sand. It is thought that accidents and predation meant that only a handful of *Argentinosaurus* ever survived through to adulthood.

Fossilized egg sites from dinosaurs closely related to *Argentinosaurus* indicate that at certain times of the year many hundreds of adults would gather together in order to nest. They seem to have preferred wide, flat floodplains, perhaps because there was the space to accommodate these large animals. Once there, the dinosaurs would have laid 3–13 eggs, each around 22 cm (9 in) in diameter, and then buried them in sandy soil. After hatching, the juvenile *Argentinosaurus* would have grown quickly, reaching adulthood in around 15 years, although it would have taken many more years' growth for the largest individuals to reach their maximum weight. Considering that *Argentinosaurus* lived on a diet of relatively non-nutritious plant material, it had a very high growth rate indeed. How large dinosaurs achieved this remains a mystery.

Fascinating Fact > Between hatching and adulthood an *Argentinosaurus* would need to grow by a factor of 25,000.

The first *Argentinosaurus* remains were discovered in 1987 by an Argentinian rancher, who initially mistook them for a piece of old wood. A complete skeleton has yet to be found, but its overall size can be estimated from those bones already to hand: for example, individual backbones are each nearly as tall as a man.

As the largest-known land animal, *Argentinosaurus* has raised questions as to just how large it is physically possible for an animal to get. Mammals cannot grow too large without their warm-bloodedness causing them to overheat; the more cold-blooded dinosaurs did not suffer from this problem, but the largest ones would have had problems moving about and, more importantly, finding enough food to eat. It is unlikely that a dinosaur could have grown much larger and heavier than *Argentinosaurus* without encountering these problems.

Argentinosaurus was an early member of the titanosaurids, a subgroup within the sauropod group of dinosaurs (see *Diplodocus*). In comparison to the other main groups of sauropod – the diplodocids (e.g. *Diplodocus*) and the brachiosaurids (e.g. *Brachiosaurus*) – the titanosaurids evolved quite late; the oldest (found in Africa) date from around 150 million years ago. Within a short time the titanosaurids had diversified and spread themselves across the world. They were especially common in the Late Cretaceous, and their remains from this period have been found on all the continents except Antarctica. They became extinct only 65 million years ago, at the end of the Cretaceous period.

Top > Injured or elderly *Argentinosaurus*, like this one, were vulnerable to attack by large predatory dinosaurs such as *Giganotosaurus*.

Pteranodon
A famous crested pterosaur

name	*Pteranodon* (teh-RAN-o-DON), meaning 'toothless wing'	**size**	9-m (30-ft) wingspan
animal type	Pterosaur reptile	**diet**	Carnivorous
lived	120–65 million years ago	**fossil finds**	North America, South America, Europe, Asia

Pteranodon was a large, graceful pterosaur that was designed to live in the air. Its vast wings and light-weight frame meant that it could use thermals (rising air currents) to soar effortlessly for hours. However, *Pteranodon* was ungainly and awkward on land, putting it at risk from predators. For this reason, it probably landed only to mate, nest or rest.

Although *Pteranodon* had a wingspan of 9 m (30 ft), its body weight was only around 20 kg (44 lb). Fossils of this animal have been found in areas that were over 160 km (100 miles) from the coast, which means that it could cover long distances, probably by gliding, which expends only a minimal amount of energy. It would have spent much of its time circling estuaries, lakes and seas, periodically swooping down to grab a fish and swallow it whole. Any *Pteranodon* that found itself unable to take off through illness or injury would have been doomed either to starve or to be caught and eaten by a predator, such as **Tyrannosaurus**.

The out-sized head crest on *Pteranodon* is found only on the males, and was probably used during mating displays or to intimidate male rivals. *Pteranodon* would have laid eggs and made nests, much as many modern seabirds do. They would also have needed to provide food for their hatchlings, flying out to sea to catch fish, then bringing them back to the nest.

The first *Pteranodon* skeleton was discovered in Kansas in 1876, but hundreds of skeletons have since been found across the world. Some of these, however, are likely to be misidentifications of other pterosaur species. As a result, there is a debate as to how many species of *Pteranodon* there actually are.

Many theories surround the purpose of *Pteranodon*'s extraordinary head crest, which could be up to 90 cm (3 ft) long. It was thought that the crest was used to help steer while flying (a bit like a boat's rudder), but wind-tunnel tests suggest that this would have put undue stress on the neck. Life-size working models of *Pteranodon* seem to confirm that the crest was more of a hindrance than a help when flying. Given that the male's crest was bigger than the female's, it is more probable that it would have been brightly coloured and used as part of a courtship display.

Fascinating Fact > *Pteranodon*'s head crest could be almost as long as the trunk of its body.

After millions of years dominating the skies, around 70 million years ago the number of pterosaur species declined to only a handful. What caused this decline is a matter of debate, but one reason could be their increasing size. By the end of the Cretaceous period, species such as *Pteranodon* and *Quetzalcoatlus* had wingspans of 9–11 m (30–36 ft). It is an unwritten rule that larger animals are more vulnerable to extinction than smaller ones, so perhaps the large pterosaurs were victims of their own success or of a marked cooling of the climate that occurred around 70 million years ago. By the time of the extinction of the dinosaurs, 65 million years ago, there was only one known species of pterosaur on Earth – *Quetzalcoatlus*. Towards the end of the Cenozoic era the role of the pterosaurs was taken by the birds.

Therizinosaurus
The giant claw

name	Therizinosaurus (ther-IZ-in-oh-SAW-rus), meaning 'scythe reptile'	size	10–12 m (33–40 ft) long
animal type	Theropod (therizinosaur) dinosaur	diet	Herbivorous
lived	75–70 million years ago	fossil finds	Asia

With its giant claws, long neck, bulbous body and small head, *Therizinosaurus* was one of the strangest-looking of all the dinosaurs. Its appearance probably stems from its unusual lifestyle, for while it belongs to the theropod dinosaurs (usually a predatory group that includes **Tyrannosaurus**, **Velociraptor** and **Giganotosaurus**), it had forsaken meat in favour of a diet of leaves and shoots.

Using its long neck, *Therizinosaurus* could reach high into the trees. Any branches that were just out of reach could be pulled towards the mouth using its giant claws. Once inside the mouth, any foliage would be stripped by the peg-like teeth and swallowed into a large stomach in order to be digested.

The claws could be almost 1 m (3.3 feet) long, and they are the biggest of any known animal's. They had several probable uses. When mating or trying to defend itself, *Therizinosaurus* may have stood with its arms outstretched, like a swan flapping its wings, so that it could display the size of its claws. If this failed, it may have been able to take gentle swipes at any aggressors, although the weight of the claws would have limited their use as an effective weapon. When walking, *Therizinosaurus* probably folded its arms

Below > At almost 1 m (3.3 ft) long, the claws of *Therizinosaurus* are the largest-known of any animal's, yet this dinosaur was a vegetarian.

Fascinating Fact > When *Therizinosaurus* was first found, scientists mistakenly identified its gigantic claws as turtle ribs.

against its body, a bit like a bird, to stop the weight of the claws from unbalancing it.

Perhaps unsurprisingly, the first specimens of *Therizinosaurus* to be found were its giant claws. These were discovered in 1948 in the Mongolian Gobi desert. In subsequent years other pieces of its skeleton were found, but for several decades the animal was a mystery to scientists, none of whom could decide what sort of dinosaur it was, what it ate or even what it looked like. It was only in the mid-1990s that it was recognized as a member of the theropod group of dinosaurs. In 1992 a nest of eggs was discovered in China, which belonged to a dinosaur closely related to *Therizinosaurus*. The tiny fossilized embryos had near-complete skeletons, and at last allowed scientists to make an accurate reconstruction of *Therizinosaurus*.

Since then, several similar species have been found, all of which have been placed together in their own special group known as the therizinosaurs (sometimes segnosaurs). This group is still little understood, but the animals' main characteristics seem to be a small skull, long neck, large claws and, most strangely, four-toed feet (theropods usually have only three toes). They are also the only known herbivorous theropod dinosaurs. Quite why they adopted such a different way of life from other theropods is not known.

The oldest therizinosaur fossils date from around 200 million years ago, but there is then a large gap in the fossil record until around 120 million years ago, after which several species are known, the last of which died out around 70 million years ago. With the exception of one lone species and a set of four-toed footprints that occurred in North America, all the therizinosaur fossils have thus far been found in China and Mongolia. Some scientists see close similarities between the primitive, bird-like dinosaurs, such as **Ornitholestes**, and the therizinosaurs. As a result, some artistic reconstructions of *Therizinosaurus* show it as having feathers on its arms and body, although definite fossil proof of this has yet to be found.

Tarbosaurus
The Asian Tyrannosaurus

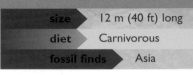

name	*Tarbosaurus* (TAR-boh-SAW-rus), meaning 'alarming reptile'	size	12 m (40 ft) long
animal type	Theropod (tyrannosaurid) dinosaur	diet	Carnivorous
lived	75–65 million years ago	fossil finds	Asia

Tarbosaurus was a large carnivorous dinosaur that ranks among the largest land predators of all time. Given that it is a very close relative of **Tyrannosaurus**, this is perhaps not all that surprising. Considering its size, *Tarbosaurus* was actually a lightly built and relatively nimble animal, with long, powerful legs that would have allowed it to move at speed. It may have hunted dinosaurs that were almost as large as itself, and probably needed to eat once every two or three weeks.

The range of habitats that *Tarbosaurus* lived in was quite wide. Its fossils have been found in areas that would have been forested, as well as in those that would have been desert or semi-desert. The sense of smell in *Tarbosaurus* was good and may have played a role in locating its food. It would have chased its prey, trying to exhaust it or injure it. Its powerful jaws and sharp teeth could rip through bone and flesh, inflicting fatal wounds. Death would have been swift. *Tarbosaurus* would then dissect the corpse, removing as much meat as possible, before leaving the body to other scavengers.

The first *Tarbosaurus* fossils were found in China in 1955, and it is now known from several specimens discovered there and in neighbouring Mongolia. It is part of the tyrannosaurid group of theropod dinosaurs, and its skeleton is nearly identical to that of **Tyrannosaurus**. This has led to arguments as to whether *Tarbosaurus* should be renamed as a new species of **Tyrannosaurus**. Despite some scientists nick-naming *Tarbosaurus* 'the Asian T-Rex', there are sufficient differences between the skulls of these two animals to warrant giving them two separate names.

Mononykus
A strange-clawed dinosaur

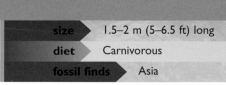

name	Mononykus (MON-oh-NY-kus), meaning 'single claw'		size	1.5–2 m (5–6.5 ft) long
animal type	Theropod dinosaur		diet	Carnivorous
lived	80–70 million years ago		fossil finds	Asia

Mononykus was a small dinosaur with long, skinny legs. It moved about on two of them, was very nimble and could run at high speeds, something that would have been useful in the open desert plains where it lived. It had a small skull, and its teeth were small and pointed, suggesting that it ate insects and small animals, such as lizards and mammals. Large eyes allowed *Mononykus* to hunt by night, when it was cooler and there would have been fewer predators (such as **Velociraptor**) about.

The biggest mystery surrounding *Mononykus* is the two single claws on its front limbs. These had powerful muscles attached to them that would have made them very strong, but even so, scientists have not been able to find a practical use for these claws. They were too short to be of use in defence or for feeding, so perhaps they were used as part of a mating display. Given the strong muscles behind them, even this theory has its problems. It used to be thought that the claws could have been used for digging, but quite how *Mononykus* would have got the claws near to the ground in the first place has not been explained.

The first *Mononykus* fossils were found near the famous Flaming Cliffs in Mongolia during the 1920s, but it was not until 1992 that more complete skeletal material was discovered and the animal could be named. *Mononykus* has proved to be something of a problem to science. Its skeleton is remarkably bird-like and lightly built, with long legs and a well-developed sternum (breastbone). In fact, *Mononykus* has so many bird-like qualities that there is an ongoing argument as to whether it is actually an unusual type of bird or an unusual type of theropod dinosaur.

The issue has not yet been resolved, but whether it was a bird or a dinosaur, *Mononykus* is usually reconstructed with a covering of feathers. In life these would have provided it with insulation during colder months. They could also have provided it with camouflage, or perhaps have been used as part of a colourful mating display.

Above > The swift-moving and sharp-eyed *Mononykus* was probably most active at night, when predators were scarce and insects plentiful. Some scientists think *Mononykus* may have behaved like the living cassowary bird: when confronted with danger, it would have kicked out with its clawed feet. Its long legs may have allowed it to run at high speeds, maybe even to jump short distances.

Fascinating Fact > Scientists cannot agree on whether *Mononykus* was a bird or a dinosaur.

Velociraptor
A famous little raptor

name		*Velociraptor* (vuh-LOSS-ih-RAP-tor), meaning 'swift hunter'	size		2 m (6.5 ft) long
animal type		Theropod (dromaeosaurid) dinosaur	diet		Carnivorous
lived		80–70 million years ago	fossil finds		Asia

The small, predatory dinosaur *Velociraptor* would have been a common sight in the Cretaceous deserts of Asia. It was ideally suited to hunting in the hot, sandy deserts, being small, lightweight and athletic. It would have been capable of agile running and jumping.

Velociraptor was probably a pack-hunter that would have stalked and attacked medium-sized dinosaurs, such as **Mononykus** and **Protoceratops**. It could kill larger prey by jumping at it with claws outstretched, then dragging a specially sharpened foot claw down the skin, inflicting terrible wounds. Doing this several times would incapacitate any animal, after which the pack would move in and death would have been swift. As well as hunting larger animals, individual *Velociraptor* would eat smaller animals, such as mammals, lizards and insects. Juveniles lived by raiding the eggs and hatchlings of other dinosaurs.

The first *Velociraptor* skeletons were discovered in Mongolia in the 1920s by the fossil hunter Roy Chapman Andrews. They attracted some interest at the time because of certain bird-like features, but the real scientific significance of the animal would take some decades to realize.

Velociraptor is part of the dromaeosaurid group of theropod dinosaurs, and is related to other agile killers, such as **Utahraptor**. The first skeleton of a dromaeosaurid dinosaur (*Dromaeosaurus*) was discovered in 1914 by Barnum Brown, but the group remained little understood until the discovery and study of *Deinonychus* by John Ostrom in the 1960s. Aside from clearly being efficient and deadly hunters, the dromaeosaurids also share certain features of their skeletons with primitive fossilized birds, such as *Archaeopteryx*, which helped to convince scientists that birds are descended from coelurosaur dinosaurs, which were closely related to the dromaeosaurids (see **Ornitholestes** for more details). The oldest-known dromaeosaurids date from around 135 million years ago; *Velociraptor* is one of the youngest-known species.

For decades *Velociraptor* was a dinosaur whose name was known only to scientists. Then, in 1992, it became famous after playing a starring role in the blockbuster film *Jurassic Park*. However, the *Velociraptors* in *Jurassic Park* were made to be far bigger and more intelligent than they would have been in real life. In terms of size they were more comparable to **Utahraptor**.

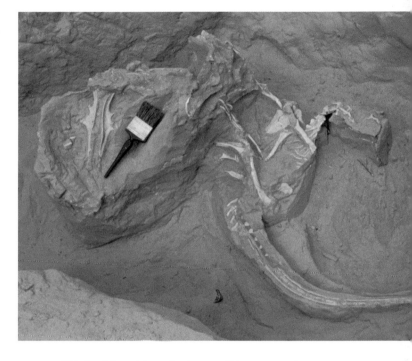

Above > This fossil from Mongolia shows a *Velociraptor* and a **Protoceratops** locked in combat; the *Velociraptor* is trying to rake its opponent's belly while the **Protoceratops** bites at its arms. This is evidence that *Velociraptor* were every bit as vicious as their cinema reputation dictates.

Many *Velociraptor* skeletons have been found, especially in Mongolia. The most famous and revealing fossil is that of the 'fighting dinosaurs', which shows a *Velociraptor* and a **Protoceratops** killed by a sandstorm while locked in combat with each other. This fossil not only provided proof that *Velociraptor* was a predator, but it is still the only known fossil that shows one dinosaur in the act of attacking another one. Another remarkable fossil find was of two juvenile *Velociraptors*, which were killed while in the act of raiding another dinosaur's nest. Some modern reconstructions show *Velociraptor* with feathers but there is no firm evidence for this yet.

Fascinating Fact > The *Velociraptors* depicted in the film *Jurassic Park* were well over twice the size of the real animal.

127

Protoceratops
A tough Mongolian herbivore

name	*Protoceratops* (pro-toh-SAIR-a-tops), meaning 'first horned face'	size	2 m (6.5 ft) long
animal type	Ornithopod dinosaur	diet	Herbivorous
lived	85–70 million years ago	fossil finds	Asia

Protoceratops was a small dinosaur about the size and shape of a large pig. Its stout, barrel-shaped body was very sturdy and would have provided some protection against the sun and water loss. This was useful because it lived among the hot, dry sand-dunes of the ancient Mongolian desert.

As a small herbivorous dinosaur, *Protoceratops* lived in loose herds that would have wandered slowly across the landscape, randomly searching for food. Some parts of the terrain in which it lived were dominated by shifting sands in which plants could be rare. In these conditions *Protoceratops* would have had a hard time finding fresh leaves and shoots, so may have fed on buried roots and tubers, digging them up and then using its sharp beak to slice through their toughened exterior. Finding water would also have been a problem, although much of this would have come from the plants it ate.

Owing to its small size and abundance, *Protoceratops* was near the bottom of the food chain, and it was hunted by many species of predatory dinosaur, including **Tarbosaurus** and **Velociraptor**. Living in herds, plus its stout frame, provided some defence against other dinosaurs, but fossils such as the famous 'fighting dinosaurs' (see **Velociraptor**) show that *Protoceratops* was regularly preyed upon.

Dozens of complete *Protoceratops* skeletons have been recovered from Mongolia and China. Many of them seem to have died after being engulfed by shifting sand-dunes; some tried to 'swim' their way out of trouble before being overwhelmed. Complete nests of dinosaur eggs found in Mongolia have been attributed to *Protoceratops*, although some have since been discovered to belong to other dinosaurs.

Protoceratops was a ceratopsid dinosaur, which means that it is related to horned dinosaurs, such as **Torosaurus** and *Triceratops*. However, *Protoceratops* is much smaller than these, has only a small head crest and no horns, so is considered to belong to the 'proto-ceratopsids', a primitive branch within the larger ceratopsid group.

The precise origin of the ceratopsids is uncertain, but a recently discovered 125-million-year-old fossil from China suggests that they may have evolved from a primitive group of ornithopod dinosaurs called the heterodontosaurids (see **Iguanodon**). Many of the earliest ceratopsids walked on two legs, with some, such as *Psittacosaurus*, possessing a thick, dome-shaped skull that was probably used for head-butting other dinosaurs.

During the Early Cretaceous period (144–99 million years ago) the ceratopsids increased in number and began to spread across the Asian continent. By the Late Cretaceous they had established themselves in North America and had become widespread there. They may even have made it as far as South America. The ceratopsids survived through to the very end of the Cretaceous period, and died out 65 million years ago during the great extinction event that killed all the dinosaurs.

Fascinating Fact > So common was *Protoceratops* in Mongolia 70 million years ago that it has been nicknamed the 'sheep of the Cretaceous'.

Top > Female *Protoceratops* probably nested in dense colonies, leading to squabbles and even battles with their neighbours.
Above > Seventy million years ago Mongolia was covered in deserts, which, amazingly, were able to support a wide variety of animal life.

Archelon
A giant among turtles

name	Archelon (ARK-eh-lon), meaning 'large turtle'		size	4.5 m (15 ft) long
animal type	Testudine (protostegid) reptile		diet	Omnivorous
lived	75–65 million years ago		fossil finds	North America

The giant turtle *Archelon* could reach the size of a speedboat and lived in the shallow seas that covered North America 70 million years ago. Most of its food was near the surface of the water, so it would rarely have needed to dive very deep. Even though it wasn't the fastest predator in the sea, the razor-sharp beak could cut its way through shell or bone, allowing it to feed on **ammonites**, jellyfish and small fish.

Archelon's biology was similar to that of modern turtles, but on a much larger scale. The gigantic flippers would have propelled it effortlessly for hours on end, and much of its time was spent in the open sea. In fact, *Archelon* would venture on to land only to lay eggs. This would have been quite an effort for an animal of that size, and would almost certainly have been done at night to minimize the danger from land predators, such as dinosaurs.

As *Archelon* swam, it would have been accompanied by an armada of juvenile fish sheltering in the safety of its shadow, while the large shell itself would have been home to barnacles and parasites.

In order to minimize weight, *Archelon*'s shell was not solid. Instead it was made of a series of thin, parallel ribs, with leathery skin stretched across it. Despite its enormous size, *Archelon* was relatively defenceless: its head and flippers could not be withdrawn, so were vulnerable to attacks from large mosasaurs, such as ***Tylosaurus***. *Archelon* could withstand the loss of one flipper, but anything more serious than this would spell its doom.

The first *Archelon* fossil was found in 1895 from the Pierre Shale of North America. The discovery of complete skeletons has led to the suggestion that *Archelon* may have hibernated by burying itself in the sea floor.

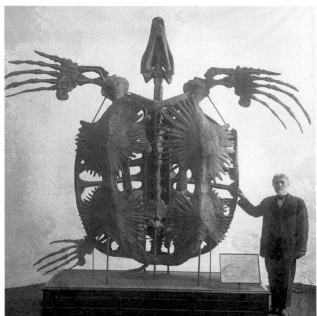

Left > Several fossils of *Archelon* have been found in the United States but many (including this one) are missing one or more of their flippers, something that is often blamed on predators such as ***Tylosaurus***. The completeness of some fossils has led to suggestions that *Archelon* could hibernate on the sea floor for long periods of time

Fascinating Fact > Some *Archelon* lived to over 100 years old. Many would take naps on the seabed.

130

Elasmosaurus
A very long-necked plesiosaur

name	*Elasmosaurus* (eh-LAZZ-mo-SAW-rus), meaning 'thin-plated reptile'	size	15 m (49 ft) long
animal type	Plesiosaur reptile	diet	Carnivorous
lived	85–65 million years ago	fossil finds	USA, Russia and Japan

Although *Elasmosaurus* could reach 15 m (49 ft) long, most of this was neck and tail. Its body was comparatively small and barrel-like, with four diamond-shaped flippers. It lived in the cool coastal waters of the northern United States, and only occasionally strayed into the hot tropical waters of the south.

It was a graceful swimmer, but could move only with its neck straight out in front; any bends or kinks in it would affect its ability to steer. When hunting, *Elasmosaurus* would place its head underneath a shoal of fish. With its body being so far away, the fish would not be aware of any danger, but a quick upwards jerk of its head would be enough to secure a catch.

Periodically, *Elasmosaurus* would venture into very shallow water. Here it would dive to the seabed and use its mouth to select and swallow smooth pebbles. Once in the stomach, the pebbles were used for ballast and for breaking up food. One fossil had over 250 pebbles in its stomach. A study of these pebbles suggested that the *Elasmosaurus* may have travelled thou-

Fascinating Fact > We humans have seven vertebrae in our necks; *Elasmosaurus* had 75.

sands of kilometres during its lifetime, picking up stones from different parts of the coast. As it was far too large and cumbersome to leave the water, it seems likely that, like ichthyosaurs, it would have given birth to live young at sea.

The first *Elasmosaurus* skeleton was discovered in 1867 and given to palaeontologist Edward Cope, who famously put the skull on the end of the tail by mistake. The animal was long believed to be able to venture ashore, but a recent study shows that it would not have had the strength to support its neck out of the water. *Elasmosaurus* was a long-necked plesiosaur (like **Cryptoclidus**). It was one of last plesiosaurs to evolve, and became extinct around 65 million years ago, just before the death of the last dinosaurs.

131

Tylosaurus
A monstrous marine lizard

name	*Tylosaurus* (TIE-lo-SAW-rus), meaning 'swollen reptile'	size	15 m (49 ft) long
animal type	Squamate (mosasaur) reptile	diet	Carnivorous
lived	89–65 million years ago	fossil finds	North America and Europe

Tylosaurus was a giant predator, whose snake-like form could be found in shallow seas across the world. It belonged to the mosasaurs, a group of marine reptiles that were the top predators in the Late Cretaceous seas. Masters of their environment, the mosasaurs ate slow-moving animals, such as **ammonites**, birds and turtles, but they would also tackle larger and swifter prey, such as sharks and plesiosaurs.

The mosasaurs were not fast swimmers. Instead of chasing their prey, they stalked it, using the natural cover provided by seaweed and rocks, making a sudden burst of speed at the last minute. Being caught in the grip of a mosasaur's jaws meant almost certain death. Some mosasaur skeletons have been found with the remains of 6-m (19.5-ft) sharks inside them, as well the large bird *Hesperornis*.

Not all mosasaurs were giants or vicious killers. There were small species, such as *Carinodens*, which grew to only 2 m (6.5 ft) in length, and some ate sea shells. The small mosasaurs were vulnerable to attack from larger mosasaurs and sharks; one fossil mosasaur had over 2000 sharks' teeth embedded in it.

Above right > A fearsome battery of teeth and powerful jaws made monosaurs the top predators in the Late Cretaceous seas.
Below > *Halisaurus* was much smaller than *Tylosaurus*; it hunted in the shallow seas surrounding the newly formed Atlantic Ocean.

For many years mosasaurs were thought to lay eggs, but in 1996 a skeleton was discovered that had two baby mosasaur skeletons inside it, suggesting that they gave birth to live young. Some juvenile mosasaurs have been found in rocks that were laid down many kilometres from the nearest coastline; these small animals would have had trouble fending for themselves, so it is thought that an adult would have needed to be nearby to protect and feed them. This suggests that mosasaurs probably nursed their young for a few weeks after giving birth.

The first mosasaur fossil was discovered in 1780 in a mine close to the Dutch town of Maastricht. The fossil was later captured by an invading French army and taken to Paris, where it was studied by the famous anatomist Georges Cuvier.

Mosasaurs adapted to life in the sea around 95 million years ago. They are thought to have evolved from a group of small, lizard-like animals called aigialosaurs that lived in coastal areas during the Late Jurassic and Early Cretaceous periods. Both the aigialosaurs and the mosasaurs were part of the reptile group squamata. This means that they share a common ancestor with living lizards and snakes. The earliest mosasaurs were quite small; it was only towards the end of the Cretaceous that several species reached gigantic proportions. The larger mosasaurs were found worldwide and were the top predators in the seas until their extinction around 65 million years ago.

Fascinating Fact > Mosasaurs were more closely related to modern lizards than the other prehistoric marine reptiles.

Xiphactinus
An ugly Cretaceous fish

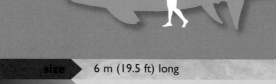

name	*Xiphactinus* (zee-FAK-tin-us), meaning 'swift swordfish'		size	6 m (19.5 ft) long
animal type	Teleost fish		diet	Carnivorous
lived	87–65 million years ago		fossil finds	North America

Xiphactinus was one of the swiftest predators in the Cretaceous seas. It was also one of the ugliest, and has been nicknamed the 'bulldog fish' because of its compressed face. Despite its unpleasant appearance, *Xiphactinus* was a swift and powerful fish. Its streamlined body and broad, muscular tail could propel it at speeds of up to 60 kph (37.5 mph) – easily fast enough to outrun almost all the other fish around it.

Much of *Xiphactinus*'s time would be spent cruising the surface waters of the shallow seas that once covered parts of North America, hunting for fish. Its lower jaw was specially hinged so that its mouth could be made particularly wide. When in pursuit of prey, it would accelerate rapidly for the kill and swallow its victim whole. As well as fish, *Xiphactinus* hunted seabirds, such as **Hesperornis**, and possibly even pterosaurs.

Fascinating Fact > Like a modern dolphin, *Xiphactinus* could jump clear of the water.

When moving at high speed, *Xiphactinus* was capable of leaping clear of the water. Splashing back into the ocean would have dislodged any loose scales or attached parasites. However, *Xiphactinus* was vulnerable to attack; its remains have been found inside the stomachs of large sharks.

The first *Xiphactinus* fossil was found during the 1850s in Kansas. Since then, many specimens have been found, including one 5 m (16.5 ft) long, containing a complete fish 2 m (6.5 ft) long that had been swallowed head first.

Xiphactinus was a teleost (bony) fish, and thus related to the majority of living fish species on the Earth today. It was the largest of several similar predatory species that lived during the Cretaceous period, none of which survived the extinction event of 65 million years ago.

Left > This *Xiphactinus* was probably killed by the 2-m (6.5-ft) long fish (still visible in its gut) that it swallowed head first.

Hesperornis
A giant diving seabird

name	*Hesperornis* (HES-per-OR-nis), meaning 'western bird'
animal type	Hesperornithiform bird
lived	80–65 million years ago

size	2 m (6.5 ft) tall
diet	Carnivorous
fossil finds	USA and Canada

Hesperornis was a large, flightless bird that was as long as a man is tall. It spent most of its life floating on the sea's surface, occasionally ducking underneath the water in pursuit of shoals of small fish. Its sleek body, long kicking legs and webbed feet would have propelled it through the water at high speed, while the small wings steered and stabilized it. *Hesperornis* could travel great distances, drifting or swimming many kilometres offshore.

When mating and nesting, *Hesperornis* may have gathered in large numbers on rocky reefs and offshore islands. Getting out of the water and on to land was quite a struggle, as the bird's long, thin legs were not strong enough to support its body weight. So instead of walking, *Hesperornis* had to lie on its belly and push itself forwards, using its feet.

On land the slow-moving *Hesperornis* was vulnerable to attack by dinosaurs and nest-raiding pterosaurs, but it was not a great deal safer when out at sea. The remains of *Hesperornis* have been found inside a range of large marine predators, including some species of shark and the mosasaur ***Tylosaurus***. It would have been especially vulnerable when resting on the

Fascinating Fact > In the 1890s a scientific study of *Hesperornis* cost so much money that the US government sacked the scientist concerned.

surface because a predator could easily grab it from underneath.

The first *Hesperornis* bones were found in 1870 by the palaeontologist Othniel Marsh in western Kansas. He was shocked to find that *Hesperornis* had teeth, a fact that provided him with strong evidence that the birds were descended from meat-eating theropod dinosaurs. *Hesperornis* has been placed in its own subdivision within the birds (the hesperornithiforms), but there are many other types of toothed bird found in the Cretaceous of North America, China and North Korea. The presence of teeth in a bird is a very primitive characteristic that means the toothed birds were an evolutionary offshoot from the toothless birds that are alive on Earth today. By 65 million years ago all the toothed birds were extinct. They were possibly victims of the same event that killed off the dinosaurs.

Tyrannosaurus
The most famous of dinosaurs

name	Tyrannosaurus (tye-RAN-uh-SAW-rus), meaning 'terrible reptile'	size	12 m (40 ft) long
animal type	Theropod (tyrannosaurid) dinosaur	diet	Carnivorous
lived	75–65 million years ago	fossil finds	USA and Canada

Tyrannosaurus is arguably the most famous and influential dinosaur of all time. It was a massive, bipedal dinosaur with a powerful tail, large head and tiny arms. An inhabitant of the dry open plains of North America, it would roam widely in search of prey – mostly other dinosaurs. It had a superb sense of smell, which may have been used to find a mate or possibly to locate dead bodies from which it could scavenge meat.

The primary weapon possessed by *Tyrannosaurus* was its mouth and teeth. Its jaws could be 1.2 m (4 ft) long, with a gape 1 m (3.3 ft) wide. The many curved and serrated teeth, most of them longer than a human hand, would be used to attack an animal, holding it in a vice-like grip. In the process, bones would be broken, arteries punctured and major organs damaged, so it did not take long for prey to be fatally injured. *Tyrannosaurus* could then devour its victim by tearing off the flesh.

Being unable to chew, it had to swallow its food whole, and could probably gulp up to 70 kg (154 lb) of meat in one go. However, there were dangers in doing this: a fossil of a large dinosaur related to *Tyrannosaurus* shows that it died of choking after two bones got stuck in its throat.

Considering the size of its body, *Tyrannosaurus*'s tiny, two-fingered arms were very small, and their use has always puzzled scientists. They were far too short to reach its mouth, and too weak to be much use in defence or hunting, but they may well have acted as a counterbalance to its gigantic head, or they could have been used as part of a mating ritual. Tooth-marks left in fossils show that individual *Tyrannosaurus* would occasionally fight one another, possibly in disputes over territorial rights.

Above > *Tyrannosaurus* possessed one of the most robust skulls of any dinosaur. A recently discovered specimen shows tooth-marks on the snout where it had been attacked by another *Tyrannosaurus*; the same skeleton also had a tiny third finger which was so small that in life it would not have been visible.

Fascinating Fact > It has been estimated that *Tyrannosaurus rex* would have needed to eat the equivalent of 265 people a year in order to stay alive.

The first *Tyrannosaurus* skeleton was discovered by palaeontologist Barnum Brown in Hell Creek, Montana, in 1902. Bones from around 20 skeletons have been found in Canada and the USA. The most complete *Tyrannosaurus* skeleton is nicknamed 'Sue'. She became the subject of a bitter custody battle between its discoverer and the owner of the land she was found on. Sue was eventually sold for over $7 million, and is now on display in Chicago's Field Museum.

Tyrannosaurus is one of several related species that are collectively referred to as tyrannosaurids (see ***Tarbosaurus***). All the tyrannosaurids were predatory theropod dinosaurs, but many were small in comparison to *Tyrannosaurus*, being only a few metres long. The oldest known tyrannosaurid fossils date from around 150 million years ago, but it was not until about 99 million years ago (the Late Cretaceous period) that they started to diversify and spread themselves across the world. Despite their relative success, none of the tyrannosaurids survived the great extinction event that engulfed the world 65 million years ago. *Tyrannosaurus* was therefore one of the last dinosaurs on Earth.

Tyrannosaurus is no longer known as the biggest of the meat-eating dinosaurs; that crown has been taken by ***Giganotosaurus*** from South America. It is, however, easily the most famous dinosaur, and has been the subject of countless books, films and television programmes. Its familiar shape makes it one of the icons of science, and it is considered by many scientists to be the most important fossil of all time.

Torosaurus
A magnificent crested dinosaur

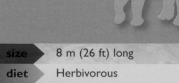

name	Torosaurus (TOR-oh-SAW-rus), meaning 'bull reptile'		size	8 m (26 ft) long
animal type	Ornithischian dinosaur		diet	Herbivorous
lived	71–65 million years ago		fossil finds	North America

Torosaurus was the largest of the horned cerotopid dinosaurs. It was a herbivore and had the most powerful jaw muscles of any known dinosaur. These, when combined with its sharp beak and a battery of around 600 teeth, allowed *Torosaurus* to slice and chew its way through almost any type of plant, including tree branches.

The body of *Torosaurus* was large, powerful and, being low to the ground, extremely stable, but it was heavy and its legs were not built for speed. This made it vulnerable to attack by the many large predators that roamed the plains, including **Tyrannosaurus**. It used to be thought that *Torosaurus* defended itself against predators using its gigantic crested skull and protruding horns. However, the skull's crest had two large, skin-covered holes in it that would have been easily damaged. Instead of defending itself with its horns, *Torosaurus* probably got most of its protection from its gigantic size and from living together in herds.

Rather than being used for defence, the head crest and horns probably played a part in mating. Male *Torosaurus* had a larger crest and horns than the females. A network of blood vessels allowed the animal

Above > Given the fragility of *Torosaurus*'s horns, mating battles would have been more about bravado and posturing than trying to inflict serious damage.

Below > *Torosaurus* lived at a time when climate change saw a marked cooling in global temperatures and a drop in sea levels. The impact of an asteroid, 65 million years ago, was the final *coup de grâce* for the dinosaurs.

to flush its crest with blood, perhaps creating a colourful display that would help to attract a mate and/or ward off rivals. Evenly matched male *Torosaurus* would probably have fought each other, charging and interlocking horns until one submitted. The horns were fragile and would have been easily chipped in battle.

The first *Torosaurus* fossils were discovered in 1891 in Niobara County, Wyoming. So far, the animal is known from only two skull specimens, both of which were found in the same quarry. One of these skulls shows evidence of a disease that is thought to be a form of bone marrow cancer, although whether it was this that actually killed it is not known.

Torosaurus is part of the ceratopid group of dinosaurs, which is itself part of the wider ornithischian (bird-hipped) dinosaurs. The ceratopids were characterized by their low, round bodies and thick (usually horned) skulls, but beyond this there were several subgroups; the two largest subgroups were the short-

frilled ceratopids and the long-frilled ones. The short-frilled species includes the famous *Triceratops*, which had a relatively small head crest that did not stretch much above its neckline. The long-frilled ceratopids, which include *Torosaurus*, had crests that could be more than 1 m (3.3 ft) in length. The difference in head crest length is thought to reflect differing lifestyles between these two types of horned dinosaur. In addition to these subgroups were the protoceratopids, characterized by **Protoceratops**.

Torosaurus lived right at the end of the Cretaceous period and was one of the last dinosaurs on Earth. It was probably a casualty of the great extinction event of 65 million years ago.

Fascinating Fact > The skull of *Torosaurus* measured 2.6 m (8.5 ft) – the longest skull of any land animal.

Ankylosaurus
The ultimate armoured dinosaur

name	Ankylosaurus (an-KILE-oh-SAW-rus), meaning 'stiff reptile'
animal type	Ornithischian (ankylosaurid) dinosaur
lived	71–65 million years ago

size	9 m (30 ft) long
diet	Herbivorous
fossil finds	North America

Ankylosaurus was a heavily armoured giant of a dinosaur, with a plated back, a club tail and a thick skull. It was herbivorous, and had a sharp beak that could slice through thick plant stems. However, its teeth were small and feeble, and could not break down food by chewing or crushing, so much of its diet was swallowed whole. Owing to its weight and armour-plating, *Ankylosaurus* was unable to lift its head up high, so it was forced to browse on low plants and bushes.

There's no doubt that *Ankylosaurus* was one of the best defended of all the dinosaurs. Across its back, tail and skull were many large, bony plates that locked together to form a protective shield that was virtually impenetrable to the teeth and horns of other dinosaurs. Some people have compared this armour to that of a giant tortoise, but *Ankylosaurus* was altogether a rather more agile creature.

If attacked, *Ankylosaurus* could use its club-like tail as a weapon, swinging it from side to side like a knight's mace. The club itself was a weighty mass of fused bony tissue that was quite capable of smashing through the leg bones of even the largest dinosaurs, including **Tyrannosaurus**. *Ankylosaurus* had a set of exceptionally powerful hind-limb and tail muscles that could be used to swing its club with terrifying speed and accuracy.

It is thought that *Ankylosaurus* was a solitary animal and that it would cover long distances in search of food. Full-grown adults had little to fear from other dinosaurs, and it is possible that they were placid animals that kept themselves to themselves unless deliberately threatened or provoked. It has been suggested that males could have used their deadly club tails to fight each other in mating battles (like the extinct mammal **Doedicurus**). If this were the case, we would expect to find damage to their armour and skeletons; this is so far missing.

Top > An *Ankylosaurus* defends itself against a **Tyrannosaurus**; its club tail was capable of breaking the legs of even the largest predatory dinosaurs.

140

The first *Ankylosaurus* fossils were found in Montana in 1908 by the palaeontologist Barnum Brown. It is now known from several incomplete skeletons from both the USA and Canada. *Ankylosaurus* was part of the ankylosaurid group of dinosaurs, and was therefore also an ornithischian (bird-hipped) dinosaur. It was a distant relative of **Polacanthus** but, like many of the later ankylosaurids, *Ankylosaurus* was larger, more heavily armoured, and possessed a club tail.

The exact origins of the ankylosauria within the ornithischian dinosaurs is uncertain, but the oldest known fossils date from around 165 million years ago in England. From about 125 million years ago the ankylosaurids start to become more common and widespread, with their remains being found across Europe, North America and Asia (including Australia).

Fascinating Fact > *Ankylosaurus's* skull was so heavily reinforced that there was little room for its brain.

In the Late Cretaceous the large, well-armoured and club-tailed species (including *Ankylosaurus*) started to occur, although these seem to have been restricted to the western part of North America and the eastern part of Asia. Despite being at their most successful in the latter part of the Cretaceous, the ankylosaurids went extinct 65 million years ago, along with the rest of the dinosaurs. Fossils of *Ankylosaurus* have been found in rocks that were laid down just prior to this mass extinction, making it one of the last-known dinosaurs on Earth.

Anatotitan
The last of the duck-bill dinosaurs

name	*Anatotitan* (AN-at-oh-tie-tan), meaning 'large duck'		size	12 m (40 ft) long
animal type	Ornithischian (hadrosaurid) dinosaur		**diet**	Herbivorous
lived	71–65 million years ago		**fossil finds**	North America

Anatotitan was a large, duck-billed dinosaur that roamed the coastal regions and lowland floodplains of the Late Cretaceous. Fossils of its stomach contents show that it was herbivorous and that it ate from a wide range of trees, including conifers, oaks and willows. It would walk and graze using four legs, but could rear up on to its muscular hind legs to reach higher into the trees. In emergencies it could also rise on to two legs and run for a short distance.

There are no known defensive features on *Anatotitan* but it was very large so it could only have been targetedy by predatory dinosaurs, such as **Tyrannosaurus**. Many *Anatotitan* skeletons are associated with rivers and lakes, and so it could have entered the water in order to escape predators. Its ability to generate a loud call would have made attacking a herd of *Anatotitan* a daunting affair.

In 1978 the fossilized nests of dozens of duck-billed dinosaurs were discovered in Montana. Based on this, it seems dinosaurs like *Anatotitan* may have nested in vast colonies. After mating, each female would have made a circular nest mound in which she buried her eggs, then stood guard over them until they hatched. Afterwards the young dinosaurs would have been protected and fed by the mother for several weeks.

Left > Duck-bill dinosaurs are famous for their decorative nasal crests, with which it is thought they could make trumpeting noises to communicate with each other.

Fossils of *Anatotitan* were first found in the 1940s, but were not officially named until 1990. It belongs to the hadrosaurid group of dinosaurs, which are themselves a subgroup within the ornithopod dinosaurs (see **Othnielia**). The hadrosaurid are often called duck-bills because of their flat beaks. They first evolved around 115 million years ago, and quickly expanded to become one of the most common dinosaurs of the Late Cretaceous, before going completely extinct 65 million years ago. Given their abundance, the hadrosaurids must have formed part of the staple diet of many predatory dinosaurs.

Many excellently preserved hadrosaurid remains have been found, including ones with fossilized skin and stomach contents. In 2000 the fossil of a hadrosaurid named *Thescelosaurus* was found to have a mammal-like, four-chambered heart; until then, dinosaurs were thought to have had only three-chambered hearts. If true, this could mean that some types of dinosaur were capable of being much more athletic than was previously assumed. It has even been suggested they could have been warm-blooded, like birds and mammals.

Didelphodon
A mammal among dinosaurs

name	Didelphodon (die-DELF-oh-don)
animal type	Marsupial mammal
lived	71–65 million years ago

size	1 m (3.3 ft) long
diet	Omnivorous
fossil finds	North America

Didelphodon was a small, hairy mammal that lived at the end of the era of dinosaurs. Although only the size of a badger, *Didelphodon* was large compared to most mammals in the Mesozoic era (248–65 million years ago). It had a covering of fur, good senses of smell and sight, and was active mostly at night when there were fewer predators around. It lived in forested regions, where it would dig, and reside in, shallow burrows.

The diet of *Didelphodon* incorporated a wide range of foods, including small animals, eggs, insects and even plants. Its teeth were specially designed for crushing, so it might have been able to crunch bones and shellfish. Being warm-blooded, *Didelphodon* would have used a lot of energy to regulate its body temperature; this meant that it had to eat regularly.

Fossils of *Didelphodon* have been found in Montana. It was a mammal, and thus part of the same broad animal group that contains most of the large animal species alive on Earth today, including ourselves. In Mesozoic times the mammals were mostly very small with delicate skeletons, which means that their fossils are rarely found. The oldest mammalian fossils are found at the start of the Jurassic period (around 205 million years ago); they probably evolved from mammal-like reptiles (therapsids), such as the cynodont ***Thrinaxodon***.

It was thought that for millions of years the mammals remained an unimportant group of small, shrew-like animals that lived by foraging. However, dog-sized mammals from the Cretaceous have recently been found in China, one of which shows evidence that it had been feasting on small dinosaurs. Little is known about the mammals' early fossil record, but some time around 145 million years ago the first monotreme mammals evolved. The monotremes are today represented by the duck-billed platypus and echidna, both of which lay eggs, just like their reptilian ancestors.

About 15 million years later the marsupial mammals (of which *Didelphodon* is an example) evolved, and this group is represented today by many living species, including the kangaroo and koala bear. The first placental mammals also date from around the same time. The majority of living mammals belong to the placentals (see ***Leptictidium***), although, like all mammals, they would have to wait for the extinction of the dinosaurs before their ambitions could be realized.

Part Three

The Age of Beasts

The Cenozoic Era

The mass extinction event of 65 million years ago marks the start of the Cenozoic or 'recent' era, sometimes referred to as the Tertiary (third) era. The extinction event killed any animal that was larger than a crocodile, including all the dinosaurs. Those small animals that did survive through to the Cenozoic era found themselves in a very different world. This era is also marked by the increasing separation of the continents and the formation of their own unique plants and animals.

Palaeocene epoch (65–55 million years ago)

The Palaeocene epoch began with a devastated world, but the plants were quick to recover. Within a few hundred thousand years there were thick jungles and swamps covering much of the world; even the polar regions were covered in dense forests. Animals that survived the extinction event remained small so that they could move among the trees. The largest animals in this period were the birds. The fearsome *Gastornis*, 2.2 m (7 ft) tall, was the top predator of this age, hunting in the jungles of Europe and North America.

The extinction of the dinosaurs saw the mammals begin to expand and move into new environmental niches. Then, at the end of the Palaeocene epoch, around 55 million years ago, there was an explosion in mammalian variety. The ancestors of many modern mammalian groups appeared for the first time, including all the hoofed animals, elephants, rodents, primates, bats, early whales and sea cows. The mammals were unwittingly beginning to dominate the planet.

Above > Sixty million years ago the majority of the world's oceans were directly connected to each another via a network of narrow seaways.
Right > The tropical coastlines of the Eocene were covered in thick mangrove forests, which were ideal places for primates such as *Apidium* to live in and for small sharks to hunt in.

Eocene epoch (55–34 million years ago)

At the start of the Eocene epoch much of the Earth was still covered in thick jungle. High global temperatures created a hothouse planet. On the forest floor were primitive mammals, such as the small horse *Propalaeotherium*, and the hopping *Leptictidium*. Living in the trees was *Godinotia*, one of the first primates, while in Asia was *Ambulocetus*, a primitive whale that could walk on land.

Then, around 43 million years ago, the climate became cooler and drier. The dense jungles were replaced by woodlands and dusty plains. These more open conditions allowed the mammals to grow bigger.

Asia was home to giant brontotheres, such as *Embolotherium*, and to massive carnivores, such as *Andrewsarchus*. In the warm seas were primitive whales, such as *Basilosaurus* and *Dorudon*, while the African coast was home to *Moeritherium* and the bizarre *Arsinoitherium*.

Around 36 million years ago lonely Antarctica, stuck over the South Pole, started to freeze, causing huge ice-sheets to form over its landmass. As a consequence, the world's climate and oceans began to cool, disrupting the global weather and radically changing rainfall patterns. Many animals couldn't cope with these changes, and in only a few million years a fifth of all life on Earth became extinct. This small extinction event at the end of the Eocene is sometimes called La Grande Coupure or 'The Great Cut'.

Oligocene epoch (34–24 million years ago)

The early Oligocene had a cool, dry climate, which gave rise to wide plains, scrublands and semi-desert. The climate change at the end of the Eocene saw the extinction of many of the more ancient mammal lineages. In their place came new species, including the direct ancestors to many modern mammals, such as the rhinos, horses, pigs, camels and rabbits.

The mammals continued to produce giants. Some, such as *Indricotherium*, reached dinosaur size, while others, such as *Entelodon* and *Hyaenodon*, were terrifying hunters. There were also the first true carnivores, such as the dog-like *Cynodictis*.

As the continents continued to move about, South America and Australia became completely isolated from the rest of the world. Over the passage of time these island continents evolved their own unique fauna of marsupials and other creatures.

It was around 25 million years ago that the first grasslands began to emerge in Asia. Until then, grasses had been an insignificant part of the landscape, but from that time onwards they grew to dominate large areas of the world, eventually covering one-fifth of its surface.

Miocene epoch (24–5 million years ago)

The wet and dry seasons of the Miocene climate ensured that large areas of the planet became covered in vast grasslands. It is not easy for animals to digest grass, so herbivores had to evolve new types of teeth and digestive systems to take advantage of its abundance.

As a consequence, the grasslands were home to early species of cow, deer and horse. Many of these started to form themselves into herds that would migrate with the changing seasons. Following these herds were new types of fleet-footed predator, including cats and dogs.

Other animals preferred to keep browsing on the leaves of trees and bushes. Some grew quite large, such as the giant elephant *Deinotherium* and the strange-looking *Chalicotherium*.

The Miocene also saw the rise of mountain chains, such as the Alps, the Himalayas, the Andes and the Rockies. Some of these new mountains were high enough to disrupt the atmospheric air flow, and thus started to play a major role in global weather patterns.

Top > By 35 million years ago the globe was beginning to look more familiar. The isolation of the Antarctic continent was to have a big effect on the world's climate.
Left > A juvenile *Chalicotherium* seeks maternal protection from two *Hyaenodon*.

Pliocene epoch (5–1.8 million years ago)

In the Pliocene the world's climate became more complicated. The planet became subdivided into many climatic regions, ranging from the freezing ice-caps, through the wet temperate zones to the warmer tropics.

On every continent the open grasslands became filled with new grazing mammals and their predators. In eastern and southern Africa the dense woodlands gave way to open grasslands, encouraging the first hominids, such as ***Australopithecus afarensis***, to come down from the trees to forage on the ground.

Around 2.5 million years ago the South American continent, which had been isolated for nearly 30 million years, collided with North America. Powerful carnivores, such as ***Smilodon***, moved south into Argentina, while southern giants, such as ***Doedicurus*** and ***Phorusrhacos***, moved north into the United States in an event known as the Great Faunal Interchange.

Below > The sabre-toothed cat ***Smilodon*** was a swift and deadly predator but its dagger-like teeth – useful for killing – were a hindrance when it came to stripping flesh from prey.

Pleistocene epoch

(1.8 million–10,000 years ago)

The start of the Pleistocene saw the world plunged into the ice age. For the next 2 million years the Earth's climate would alternate between colder and warmer phases. The colder periods would last around 40,000 years, during which glaciers and ice-sheets would spread across the continents, getting as far south as London and New York. In between these cold phases would be warmer interglacial periods, when the ice would recede and the sea levels rise.

Animals such as ***Mammuthus*** (the woolly mammoth) and ***Coelodonta*** (the woolly rhino) evolved thick fur coats and a layer of fat to help them live in the colder regions. Around them were herds of deer and horses that would be hunted by predators, such as ***Panthera leo*** (the cave lion). Around 180,000 years ago these animals were also to be hunted at various times by humans, such as ***Homo neanderthalensis*** and, later on, ***Homo sapiens***.

The increasing severity of the ice age made it hard for many larger animals to cope with the extreme climate swings, causing some species to become extinct. Humans may not have helped by selectively hunting certain animals. Since the end of the last ice age, around 10,000 years ago, the world has had a warmer, wetter climate. This allowed ***Homo sapiens*** to grow in numbers and spread around the globe. Humans learnt to use the seasonal rains to their advantage and began to grow crops. Small farming communities developed into towns, and within only a few thousand years the human population had created a global society based on advanced technology. In the process, many animal species with which humans had shared the planet came under ecological pressure. As a result, we are living through, and contributing to, what palaeontologists would describe as a mass extinction event.

Top > The expansion of the ice-sheets, around 1.8 million years ago, suited some mammals. It was only when the ice started to shrink again that the problems started.
Above > During glacial phases European winters would have been harsh affairs, but through the use of technology such as fire ***Homo sapiens*** was able to survive them.

Gastornis
The bone-crushing terror bird of the Eocene jungles

name	Gastornis (gas-TOR-niss), meaning 'Gaston's bird'	size	2.2 m (7 ft) tall
animal type	Neognath (diatryma) bird	diet	Carnivorous
lived	56–41 million years ago	fossil finds	North America and Europe

Gastornis was a gigantic flightless bird with a sharp, powerful beak capable of tearing flesh and crushing bones. It was one of the first large predators to evolve after the mass extinction event 65 million years ago that heralded the death of the dinosaurs. It was carnivorous and fed mostly on small mammals, such as **Propalaeotherium** and **Leptictidium**. It is estimated that it may have weighed as much as a tonne, making it one of the heaviest birds of all time.

Being so large and heavy, *Gastornis* was not very fast-moving, but it didn't need to be. It lived in forests and swamps, where it could use the dense vegetation to stalk and ambush its prey. *Gastornis* would grab at its victim using its beak; once caught in its powerful grip, there was little hope of escape. Its beak and neck muscles were then used to crush or shake the prey to death. *Gastornis* may also have scavenged on dead bodies, and it has even been suggested that the strong beak could have been used to crack the shells of clams, nuts and seeds.

Gastornis would probably have been a solitary hunter, and may have marked out territory that it would have defended violently against intruders. The females made their nests on the ground (they were far too large to climb trees) and probably laid only a single large egg. As the nests of some modern flightless birds are guarded by both the males and females, it has been speculated that this may have been true for *Gastornis* as well. After hatching, the chick would have been helpless for a few weeks and needed the protection of its parents.

The first *Gastornis* fossils were found near Paris in 1855, but it was not until 1881 that enough bones were discovered to allow the skeleton to be reconstructed (albeit incorrectly at first). Since then, many *Gastornis* bones have been recovered from France, Germany and the USA. Some people still refer to *Gastornis* by its old name of *Diatryma*, although it is now invalid.

Left > In a world without dinosaurs the robust *Gastornis* was able to claim the prize for top predator. It would rule supreme until the evolution of larger mammalian predators such as *Hyaenodon*.

The mass extinction of 65 million years ago did not affect the birds as badly as some other animal groups. Some of the more primitive toothed birds, such as **Hesperornis**, became extinct, but more modern groups of birds (such as ducks, divers and albatrosses) survived through to the Cenozoic era, where they flourished.

Gastornis is one of several similar species that belong to the diatryma group of birds. The earliest diatryma fossils are found in 50-million-year-old rocks, and the signs are that the last of the group went extinct about 41 million years ago. They left no living descendants.

The diatrymas are part of the neognaths, a massive subdivision that contains most living species of bird. The neognaths probably evolved during the Late Cretaceous, but diversified only in the Palaeocene and Eocene epochs to produce the thousands of bird species with which we are familiar today. The exception is the ratites (large flightless birds), such as ostriches, rheas, emus and kiwis, which are part of the palaeognath birds that split off from the bird lineage in the Cretaceous, slightly earlier than the neognaths.

The decline of *Gastornis*, around 41 million years ago, coincides with a rise in the numbers of large carnivorous mammals, such as **Hyaenodon**. These may have outcompeted this slow, lumbering bird, eventually driving it to extinction.

Fascinating Fact > It has been calculated that the beak of *Gastornis* was powerful enough to crack a coconut.

Leptictidium
A hopping mammal

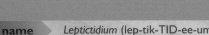

name	*Leptictidium* (lep-tik-TID-ee-um), meaning 'delicate weasel'	size	60–90 cm (2–3 ft) long
animal type	Placental (leptictid) mammal	diet	Carnivorous
lived	50–40 million years ago	fossil finds	Europe

Leptictidium was a fast-moving, medium-sized mammal that lived in the dense jungles of central Europe. It had long back legs and flattened feet that were initially a puzzle to scientists, who couldn't work out how it would have walked. It was eventually concluded that *Leptictidium* must have moved about by hopping on both legs, just like a kangaroo.

Fossils of *Leptictidium* have been so well preserved that we know exactly what it ate. It was a carnivore, preying on insects, lizards and small mammals that it probably plucked from low branches or found by rooting around the leaf litter on the forest floor. The long snout, sensitive to smell, would have twitched constantly, and had muscles that could move it in all directions, a bit like the nose on a mouse.

Leptictidium was a placental mammal, which means that its young gestated inside the female's womb, where they received nourishment from the placenta (lining). Placental mammals give birth to live young, and form the largest group of mammals on Earth today, which includes humans. The first placental mammals evolved around 130 million years ago. Before this there were several types of more primitive mammal, including the egg-laying monotremes (such as the duck-billed platypus) and the marsupials (such as kangaroos and **Didelphodon**), whose young developed inside a pouch.

The mammals seem to have survived the great extinction of 65 million years ago very well, possibly because their small size and wide diet protected them from any sudden changes in their environment. At the start of the Cenozoic the two largest mammalian groups were the marsupials and the placentals. (The third largest mammalian group, the multituberculates, declined rapidly and went extinct around 35 million years ago.)

The placental group quickly expanded into every available environmental niche, including the trees (see **Godinotia**) and the seas (see **Ambulocetus**). The marsupials were marginalized by the placentals, and became restricted to isolated parts of the world, such as South America and Australia, which is where most living marsupial species are found today.

The leptictids, the family of placental mammals to which *Leptictidium* belongs, first evolved around 75 million years ago and went extinct around 45 million years later. The best-known fossils come from the Messel Oil Shales in Germany, and are so perfectly preserved that it is possible to see their fur and internal organs.

Fascinating Fact > *Leptictidium* would have had to eat its own weight in insects every six days.

Godinotia
One of the earliest-known primates

name		*Godinotia* (god-in-OH-sha), named after the scientist Marc Godinot	size		30 cm (12 in) long without its tail
animal type		Placental (primate) mammal	diet		Insectivorous
lived		49 million years ago	fossil finds		Germany

Godinotia was a small but agile tree-dwelling primate mammal. It was smaller than a domestic cat, and probably weighed less than a bag of sugar, but nonetheless it is potentially an ancestor of ours. It is one of the earliest-known primates, the order of mammals that includes monkeys and humans.

Like all the earliest primates, *Godinotia* lived up in the trees, where food was plentiful and there were few large predators. To help it get about and feed, it had excellent binocular vision, long limbs, and grasping hands that it would have used to cling to branches or leap between trees. Its large eyes reveal that it was probably nocturnal and spent its nights hunting for small insects and ripe fruit on which it could feast.

The male and female *Godinotia* were about the same size as each other, which, if modern species are anything to go by, probably means they were solitary animals that met up only in order to mate. The large size of *Godinotia*'s penis bone indicates that mating took place over a long period of time (possibly several hours). The male did this to be sure the female would become pregnant by him. Newborn babies would be suckled by their mother for several months, and would have clung to her fur while she moved about the forest.

Superb fossils from Messel in Germany show that early primates, such as *Godinotia*, were furry and that they already exhibited many features that would help make the primates such a successful group. This includes having forward-facing eyes, dextrous fingers and an enlarged brain. The fossils also show that *Godinotia* was especially vulnerable to attack by crocodiles, which may have grabbed them from riverbanks when they came down from the trees to drink.

Godinotia is one of the oldest-known primates but the origins of this group remain controversial. Some claim that the oldest primate is 70 million years old but this is based on a single fossilized tooth. The oldest definite primate fossils date from around 55 million years ago and belong to animals that are similar to *Godinotia*.

Around 45 million years ago the primate lineage split to form firstly the monkeys, which include the apes (e.g. **Apidium**, **Australopithecus afarensis** and **Homo sapiens**), and secondly the so-called 'basal primates' (e.g. lemurs and the aye-aye). *Godinotia* was named in 2001, and so far its fossils have been found only in southern Germany.

Fascinating Fact > Owing to a large bone in its penis, *Godinotia* could mate for hours at a time.

Propalaeotherium
The cat-sized ancestor to the horse

name	Propalaeotherium (pro-pay-lee-oh-THEE-ree-um)	size	30–60 cm (1–2 ft) tall
animal type	Placental (perissodactyl) mammal	diet	Herbivorous
lived	49–43 million years ago	fossil finds	Germany

Propalaeotherium was a primitive type of horse that grew only to the size of a small dog. This would have been an advantage, as it lived in dense jungles where there was little space for large animals to move about. Unlike modern horses, *Propalaeotherium* had several toes with small hooves on the end of each – four on the front and three on the back. They then walked on the pads of their feet like cats and dogs.

These little horses were probably solitary animals that would wander the forest floor, looking for fresh leaves, flowers or fallen fruit and seeds. They were probably not territorial, and would have moved about randomly, meeting other *Propalaeotherium* only by chance. Even though *Propalaeotherium* were small, they had few predators in their habitat that would have been large enough to attack and kill them. One of the few was the giant bird **Gastornis**: the Eocene was a time when birds ate horses.

The first *Propalaeotherium* fossils were discovered in 1911 in Germany, with the best fossils coming from the Messel Oil Shales near Frankfurt. The Messel was once an ancient swamp into which animals would have accidentally stumbled. Some *Propalaeotherium* fossils are so well preserved that it is possible to see their fur, stomach contents and, in one case, even unborn foals. Over 70 fossils exist, and from them scientists have been able to deduce a great deal about the lives of these ancient animals.

Fascinating Fact > Fossils of fermented grapes were found in the stomach of a *Propalaeotherium*, which could mean that it died drunk.

Propalaeotherium belongs to the perissodactyls, a group of hoofed placental mammals characterized by having an odd number of toes on their feet. Living perissodactyls include rhinos, tapirs and horses; fossil representatives include brontotheres (e.g. **Embolotherium**), chalicotheres (e.g. **Chalicotherium** and **Ancylotherium**) and **Indricotherium**.

Related to the odd-toed perissodactyl mammals are the artiodactyl mammals, which have an even number of toes on each foot and include pigs, camels, deer and cows. Together the perissodactyl and artiodactyl mammals form the hoofed mammals or ungulates (see **Entelodon**). The early fossil history of hoofed mammals is not yet fully known, but it is thought that the perissodactyls and artiodactyls evolved from a common ancestor that lived around 60 million years ago in the early Palaeocene epoch.

The horses, the family to which *Propalaeotherium* belongs, have long been cited as a classic example of evolution in action. This is because the fossil record of horses is particularly complete, and it is possible to see the many stages that exist between *Propalaeotherium* and the modern horses of today. Over the course of 55 million years, evolution adapted the bodies of horses to the environments in which they lived. Early horses, such as *Propalaeotherium*, were small and had many toes on each foot. As time progressed and the dense forests were replaced by open plains, so the horses became bigger and the number of toes reduced. Modern horses are large animals with only a single toe on each foot, perfect for the wide grassy plains in which they live. They have adapted to eating grass, and their long legs and muscular bodies mean that they can run fast and travel long distances. *Propalaeotherium* could do none of these things.

Left > Fossils of *Propalaeotherium* from the Messel Oil Shales in Germany are so well preserved that it is possible to see their fur, unborn foals and even their last meal. It is speculated that many of the animals preserved in the Messel Oil Shales may have been suffocated by carbon dioxide fumes that were periodically released from a nearby volcanic lake.

Ambulocetus
A primitive whale that walked on land

name	Ambulocetus (amb-yoo-lo-SEE-tus), meaning 'walking whale'		size	3 m (10 ft) long
animal type	Placental (cetacean) mammal		diet	Carnivorous
lived	50–49 million years ago		fossil finds	Pakistan

At around 3 m (10 ft) long and 300 kg (660 lb) in weight, *Ambulocetus* was one of the largest mammals of its time. It was an ancestor to modern whales and a capable predator. When on land, it could walk only slowly, but was a powerful swimmer in water.

Ambulocetus lived along the estuaries, riverbanks and coasts of ancient Pakistan. It would have spent much of the day lying on the shore, basking in the sunshine. Periodically, *Ambulocetus* would enter the water and hide just beneath the surface, waiting for land animals to come close to the water's edge. When they did, it would spring upwards, grab the animal and drag it into the water, where it would be drowned and then eaten.

Ambulocetus had one intriguing adaptation to aquatic life – its hearing. It did not have the normal mammalian ears that capture vibrations in the air; instead it heard through its jawbone. This meant it could pinpoint the direction of sounds under water, but was partially deaf on land. It could, however, have sensed the vibrations of approaching animals by holding its jaw to the ground.

The life cycle of *Ambulocetus* is not well known, but evidence from minerals deposited in their teeth suggests that the females may have travelled upstream into fresh water in order to give birth. The juveniles would then grow up in the rivers, before migrating into the estuaries and coastal regions.

The first *Ambulocetus* fossils were found in 1994 in the foothills of Pakistan. Their discovery was greeted with excitement by scientists, as they helped to resolve a lengthy dispute about the evolutionary origin of the whales. It had long been suspected that whales had evolved from a group of four-legged placental mammals known as the mesonychians (of which *Andrewsarchus* is an example), but there was such a gap in the whales' fossil record that nobody could be sure.

The teeth, ears and skeleton of *Ambulocetus* show that it was definitely a primitive type of cetacean mammal (i.e. a whale), yet it also had many features of the mesonychian mammals as well. This, and the fact that it was obviously capable of moving about on land, means that it is a type of 'missing link' between the land-bound mesonychian mammals, such as *Andrewsarchus*, and fully aquatic whales, such as

Basilosaurus. It was the first decent evidence that the whales had evolved from land animals. Since then, the fossils of many other primitive whales found in the same region have filled in many more of the gaps that once existed in the whales' evolutionary history. The oldest-known fossilized whale is currently the 53.5-million-year-old *Himalayecetus*, which was found in northern India.

Below > A small horse has a lucky escape after a surprise attack by an *Ambulocetus*. Many animals looking to drink from the riverbank would have been victims of this primitive whale.

Andrewsarchus
The largest carnivorous land mammal

name	Andrewsarchus (and-rooz-ARK-uss), meaning 'Andrews' beast'		size	Up to 5.5 m (18 ft) long
animal type	Placental (mesonychian) mammal		diet	Carnivorous
lived	40–35 million years ago		fossil finds	Mongolia

Andrewsarchus was the largest meat-eating land mammal of all time. It walked on all fours, grew up to 5.5 m (18 ft) long and stood taller than a man. Its skull alone was nearly 1 m (3.3 ft) long.

Although gigantic and armed with sharp teeth, *Andrewsarchus* could not run fast, so it probably didn't actively hunt other animals. Instead, it would wander along riverbanks and lake shores in search of dead or dying animals that had been washed up by the water. With its strong teeth and jaws, *Andrewsarchus* could eat almost anything, from drowned animals to turtles and stranded fish. The more rotten the carcass, the better, but even tough objects, such as bone and shell, could be crushed to a pulp using its powerful jaw muscles.

Fossils of mesonychians, the group of mammals that includes *Andrewsarchus*, are usually found alone, so it is thought that they lived solitary lives. They may have got together only in order to mate. The animal's great size means it would have had few natural enemies.

The mesonychians were a group of placental mammals that evolved shortly after the extinction event of 65 million years ago. They were mostly small to medium-sized carnivores that were among the first large mammalian carnivores to evolve. They were never very diverse, with only a dozen species being known, but they are a very important group in the mammals' evolutionary tree.

The mesonychians looked like modern wolves and dogs, but they are actually very closely related to the hoofed (or ungulate) animals, especially the

Below > *Andrewsarchus* was a scavenger whose teeth and jaws were strong enough to crush bone and even the shells of turtles.

artiodactyls, such as pigs (e.g. **Entelodon**), deer (e.g. **Megaloceros**), camels and cows. However, the mesonychians are probably better known for being ancestors to the whales, producing species such as **Ambulocetus** and **Basilosaurus**. Although their origins are in Asia, mesonychians managed to spread themselves to Europe and North America by 50 million years ago, but eventually found themselves being outcompeted by large scavengers, such as **Hyaenodon** and **Entelodon**. The last mesonychian fossils are found around 30 million years ago.

Andrewsarchus is known from just one skull, found in the Mongolian desert in the 1920s. It was named after the famous explorer Roy Chapman Andrews, the man who is reputed to be the inspiration for the film character Indiana Jones.

Embolotherium
The first of the giant mammals

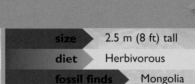

name	*Embolotherium* (em-BOL-o-THEE-ree-um), 'battering-ram beast'		size	2.5 m (8 ft) tall
animal type	Placental (perissodactyl) mammal		diet	Herbivorous
lived	40–35 million years ago		fossil finds	Mongolia

Embolotherium was a rhino-like herbivore, the largest vegetarian mammal of its time. It would have weighed around 2 tonnes, and had a large bony growth, up to 70 cm (2.4 ft) long, above its nose.

This animal lived on the open plains of Mongolia and travelled in large, slow-moving herds that would browse on trees and bushes, using their rough tongues to rip off leaves and shoots. Its stomach was not good at digesting food, so it would have to have eaten vast quantities of plant material in order to stay alive.

Being large, four-footed and sturdy, *Embolotherium* was a hardy animal with few natural predators. Its body could hold reserves of fat that would permit it to survive droughts or other extreme conditions. Even so, females probably gave birth at the start of the wet season so that there would have been plenty of fresh food for their calves.

The *Embolotherium*'s distinctive horn is made from bone covered in a thin skin layer. The male's horn was around three times longer than that of the female, but it was brittle and easily damaged, so of no real use for defence. Instead, the male *Embolotherium* probably used its horn in courtship rituals, making mock charges against its rivals, locking horns and pushing until one or other animal surrendered. Unfortunately for such a large animal, the brain was small, being only about one-third the size of a modern rhino's.

The first *Embolotherium* skeletons were discovered in Mongolia in the 1920s by palaeontologist Henry Fairfield Osborn. He recognized that it belonged to a group of placental perissodactyl mammals (see **Propalaeotherium**) known as brontotheres ('thunder-beasts'), whose fossils can be found in their thousands in some areas. For millions of years the brontotheres dominated the landscapes of Asia, North America and Europe, but by 28 million years ago they had been driven into extinction, possibly by more efficient browsers, such as **Chalicotherium** and **Indricotherium**.

Fascinating Fact > Although the male *Embolotherium* had a spectacular horn, it was brittle and probably used for display rather than fighting.

Moeritherium

A water-loving ancestor to the elephants

name	Moeritherium (mee-ri-THEER-ee-um), meaning 'beast of Moeris'	size	2 m (6.5 ft) long
animal type	Placental (proboscidian) mammal	diet	Herbivorous
lived	36–33 million years ago	fossil finds	North and West Africa

Moeritherium was about the size and shape of a very large pig. Its body was barrel-shaped, with stumpy legs, an elongated head and a long, flexible snout like that of a tapir. It was an early relative of the elephant.

Semi-aquatic by nature, *Moeritherium* lived in river estuaries and was especially fond of the shallow ponds and channels that ran through the ancient mangrove swamps that occurred along the North African coast. It probably lived in small herds that would wade their way through the water, stopping occasionally to use their long snouts and small tusks to grasp and rip up vegetation from the riverbed. It was particularly fond of seagrass, an aquatic plant that grew in profusion in the areas where *Moeritherium* lived.

The animal's rotund body and short legs made walking on land slow and ungainly, and it was not a fast swimmer either. For this reason *Moeritherium* would have been most at home when half-submerged in the water, its feet touching the bottom, its head poking out above. Living like this also made it hard for land- or sea-bound predators to attack it.

The first *Moeritherium* fossils were found in 1904 in the Egyptian desert, which, 34 million years ago, was on the coast. Other finds have been made in Libya, Mali and Senegal.

Moeritherium was a proboscidian, which means that it is part of the same group of placental mammals as the elephant. It had many elephantine features, including a trunk, tusks and very similar hearing. Even so, *Moeritherium* was not a direct ancestor to the elephant: that honour goes to *Palaeomastodon*, which lived around the same time. The proboscidians evolved in Africa and were restricted there for millions of years until, during the Miocene epoch, they were able to cross land-bridges to Europe and Asia. The later proboscidians include the woolly mammoths and the gigantic **Deinotherium**.

Fascinating Fact > *Moeritherium* lived in water, yet nowadays its fossils are most commonly found in the Sahara desert.

Arsinoitherium
A double-horned giant

name	Arsinoitherium (aars-in-oh-ith-EAR-ee-um)
animal type	Placental (proboscidian) mammal
lived	36–30 million years ago

size	1.8 m (6 ft) tall
diet	Herbivorous
fossil finds	Egypt, Oman, Libya and Angola

Despite its size and aggressive appearance, the rhino-like *Arsinoitherium* was actually a gentle giant. It lived in the North African coastal mangrove swamps, where it would have spent most of its time wallowing in the water.

Arsinoitherium would emerge on to land only for brief periods of time because its hind legs were permanently bent and pointed outwards, which was ideal for swimming but not so good for walking. Its awkward gait has been confirmed by the discovery of its fossilized footprints in Egypt. It is thought that *Arsinoitherium* would have ventured on to land only to mate or to move to new feeding areas. It had no natural predators on land or in the sea.

Although *Arsinoitherium* was vegetarian, its complex means of chewing allowed it to eat only certain types of fruit and leaf. It must therefore have spent much of its time searching for suitable quantities of food in order to maintain its large bulk.

The most obvious feature of *Arsinoitherium* is its giant double horn, which was larger in the females than the males. Given that it was hollow, and that both sexes had excellent

Fascinating Fact > *Arsinoitherium* may have used its huge double horn to make booming mating calls.

hearing, it is possible that the male used it to create a loud mating call. The males may also have used their horns in mating battles, locking them with a rival and twisting until their opponent surrendered.

Fossils of *Arsinoitherium* were first discovered in Egypt in 1902, but it took several years before enough bones were found to build an entire skeleton.

Arsinoitherium belongs to the embrithopods, a small group of placental mammals known only from a few scattered fossils found in North Africa, Turkey and Romania. Not all the embrithopods were large and horned; some were small. Although not a well-studied group of mammals, embrithopods are thought to be closely related to early proboscidians (elephant family), such as **Moeritherium**. They date from around 55 million years ago and went extinct during the Oligocene epoch (34–24 million years ago).

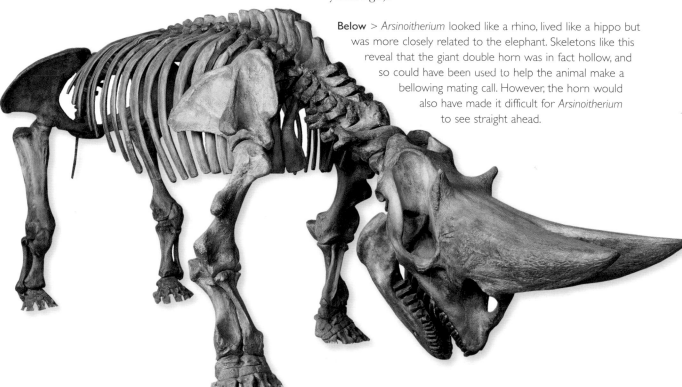

Below > *Arsinoitherium* looked like a rhino, lived like a hippo but was more closely related to the elephant. Skeletons like this reveal that the giant double horn was in fact hollow, and so could have been used to help the animal make a bellowing mating call. However, the horn would also have made it difficult for *Arsinoitherium* to see straight ahead.

Basilosaurus
The first of the giant whales

name	Basilosaurus (BASS-il-oh-SAW-russ), meaning 'regal reptile'	size	Males 21 m (69 ft) long; females 18 m (59 ft) long
animal type	Placental (cetacean) mammal	diet	Carnivorous
lived	45–36 million years ago	fossil finds	North America, Europe, Egypt, New Zealand

Basilosaurus lived in most of the warm seas around the world, and was easily the largest predator of its time. It was capable of attacking large prey, including other whales, such as **Dorudon**.

Like all whales, *Basilosaurus* was an air breather, but it had no blow-hole. Instead, it had to raise the tip of its nose out of the water to take a breath. Its ribcage was solid and not very flexible, which meant that its lung space was restricted; thus *Basilosaurus* could not stay under water for prolonged periods of time. Any attacks it made therefore had to be swift and accurate. With its high mammalian metabolism, *Basilosaurus* would have needed to eat often in order to keep its energy reserves high. It did not have the large insulating fat deposits of modern whales, so it could not have strayed into cooler waters. It was therefore restricted mainly to warmer waters, such as the ancient Tethys Sea that once ran between the African and European continents.

When it came to food, *Basilosaurus* was not a fussy eater. Fish, sharks, squid, turtles and other marine mammals could all fall prey to its keen eyesight. It would swim in pursuit of its victims, using its powerful serpentine tail to make brief but fast pursuits. Once its powerful jaws and sharp, serrated teeth clamped down on to an animal, there could be no escape.

As a probable solitary hunter, *Basilosaurus* would have spent long periods of time on its own. It lacked a 'melon', the organ that modern whales and dolphins use for echolocation and to 'sing' to one another. Even though *Basilosaurus* couldn't sing, the males and females would probably meet seasonally to breed. Given their sinuous body shape, they would have used their small back legs (a relic of their land-living ancestors) to guide one another into the correct mating position.

For many decades *Basilosaurus* was the oldest-known fossilized cetacean (or whale). Its tiny back legs were seen as proof that the whales had once been land animals, but it was not until the discovery of 'walking whales', such as **Ambulocetus**, that there was proof of this. It is now known that the whales evolved from large, wolf-like animals called mesonychians (see **Andrewsarchus**).

Above > *Basilosaurus*'s serpentine shape and lack of blubber made it look unlike any of the whales that swim in today's oceans.
Right > Speed and agility were part of *Basilosaurus*'s hunting strategy; they allowed it to hunt large prey, like this 2-m (6.5-ft) long shark.

Although it broadly resembled modern whales, *Basilosaurus* (and its close relative **Dorudon**) were evolutionary dead ends and left no ancestors. The modern whales are split in two main groups: the odontoceti (toothed whales, such as the sperm whale and dolphins) and the mysticeti (baleen whales, such as the blue whale). The earliest ancestors to both the toothed and baleen whales are found in the Early Oligocene epoch, around 30 million years ago. Since then, the whales have prospered, and are now represented by approximately 70 species, which can be found everywhere from the polar regions to the Equator. At 18 m (59 ft) long, the sperm whale is one of the largest predators of all time, while the 29-m (95-ft) blue whale is one of the largest, and certainly the heaviest, animals ever to evolve.

The first *Basilosaurus* skeleton was found in Louisiana during the early 1830s. It reminded people of a sea serpent, so was named *Basilosaurus* (regal reptile). In 1842 the mistake was realized, but the reptilian part of its name stayed in place. Its extinction came 35 million years ago, during a time of great change in the Earth's oceans. Cold Antarctic waters pushed their way north into warmer waters, rendering many marine species extinct.

Right > At 2 m (6.5 ft) long, this *Basilosaurus* skull is about the same size as an adult man.

Dorudon
A small, shark-eating whale

name	Dorudon (DOR-oh-don), meaning 'spear-toothed'		size	5 m (16.5 ft) long
animal type	Placental (cetacean) mammal		diet	Carnivorous
lived	45–36 million years ago		fossil finds	North America, Egypt

The body of *Dorudon* was streamlined, muscular and dolphin-like in shape. It would have been able to swim at high speeds and manoeuvre quickly under water, making sudden turns and even leaping clear of the sea. Its speed and agility were used for catching fast prey, such as fish, sharks and squid, and also meant that *Dorudon* could get out of the way of other predators, such **Basilosaurus**.

Dozens of *Dorudon* fossils have been found preserved together in the same geological horizons, indicating that they were social animals that lived in large groups. Even so, they did not 'sing' to one another because they lacked the melon organ that modern cetaceans use to communicate across long distances. However, it is thought that they could still have used high-pitched sounds to communicate when very close by.

When it came to breeding, pods of *Dorudon* would migrate to seasonal birthing grounds in shallow coastal regions. One such ancient breeding ground has been found in the Egyptian Fayum desert, where dozens of adult and juvenile *Dorudon* skeletons are preserved together. Some of the juvenile skulls had been crushed by the jaws of a giant predator. The only viable candidate is the larger whale **Basilosaurus**, which may have hung around the *Dorudon*, waiting for them to give birth, before attacking their newborn calves.

The first skeleton of *Dorudon* was uncovered in Egypt in 1906, but it was initially mistaken for a juvenile **Basilosaurus**. Although *Dorudon* was eventually recognized as being a separate species, the main difference between it and **Basilosaurus** is one of size. Like **Basilosaurus**, a *Dorudon* was warm-blooded, breathed air and had the same small back legs.

Apidium
One of the first monkeys

name	Apidium (ay-PID-ee-um), meaning 'small bull'	size	25–30 cm (10–12 in) long, not including the tail
animal type	Placental (primate) mammal	diet	Omnivorous
lived	36–32 million years ago	fossil finds	Egypt

Apidium was a primitive monkey that lived in the tropical forests of North Africa. Here it probably lived in small tree-dwelling troops, which travelled along branches and jumped between trees, rather like modern monkeys. It showed many new adaptations to this specialized lifestyle. Its hind feet were particularly good at grasping branches, ensuring that it didn't fall to the forest floor below, where predators might be waiting.

Unlike the nocturnal **Godinotia** (a primate ancestor), *Apidium* was a daytime feeder, using its keen eyesight to find ripe fruit and insects in the trees, which it would then eat using its specially rounded and flattened teeth. Much of its waking time would have been spent in the search for food, and it may have had to wander over a wide area to satisfy its hunger.

Male *Apidium* were bigger than the females, which, by comparing them with living primates, suggests that they probably lived in small groups, where a small number of males would have had control over several females. The males had large canine teeth, which they would use to fight one another over mating rights and for the ultimate right to control a particular group of female *Apidium*.

The first fossilized bones of *Apidium* were found in Egypt in 1907 by the members of an expedition from the American Museum of Natural History, New York. However, the first fossil was misidentified as being the jaw from a primitive cow: the mistake wasn't realized until the 1950s.

Apidium was a primate mammal and an example of a catarrhine or 'old world monkey'. The monkeys as a whole evolved from primitive primates, such as **Godinotia**, but around 45 million years ago they diverged into two groups. The platyrrhines or 'new world monkeys' are generally small, agile species that are today represented by squirrel, howler and spider monkeys, tamarins and marmosets. The catarrhines, of which *Apidium* is an early example, are larger and more diverse. Their living representatives include the apes (orang-utans, gorillas, chimps and humans), baboons, mandrills and colobines. See **Australopithecus afarensis** for more information on the evolution of the apes.

Fascinating Fact > When this monkey's bones were first found scientists thought it was a type of cow.

Hyaenodon
A swift, bone-crushing predator

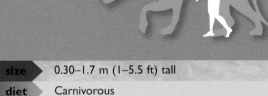

name	Hyaenodon (hi-EE-noh-don), meaning 'hyaena-toothed'	size	0.30–1.7 m (1–5.5 ft) tall
animal type	Placental (creodont) mammal	diet	Carnivorous
lived	41–25 million years ago	fossil finds	North America, Europe and Asia

Hyaenodon was a deadly predator that ranged in size from a fox to a small rhinoceros. It roamed the plains of North America, Europe and Asia, had powerful jaws, huge claws and was a fast runner. The largest species of *Hyaenodon* were the top predators of their day, and would have been feared by most other animals.

The smaller *Hyaenodon* species would have hunted at night in packs, cooperating with one another to bring down and kill prey that was larger than themselves. The biggest species of *Hyaenodon*, which stood taller than a man, probably hunted alone and may have been territorial, defending their patch against

any trespassers. They could attack and kill animals that were as big as, or larger than, themselves, such as **Chalicotherium** and **Entelodon**.

The success of *Hyaenodon* was due in part to its excellent sense of smell and acute eyesight, which would allow it to find prey or dead bodies quickly. Unlike most of the animals around it, *Hyaenodon* walked on its toes (as opposed to the soles of its feet), so it could run extremely fast. On the end of each toe was a sharp claw for killing and dismembering prey. Finally, its jaws were very strong and powerful, and could crush bone, which meant that *Hyaenodon* could get at the nutritious marrow inside.

A fossil in the USA reveals that *Hyaenodon* did not have great table manners: after making a kill, it would go around the dead body covering it in dung. This would have been done to mark out its territory and also to obscure the smell of the kill from other rival predators (some bears do this today).

It is thought that *Hyaenodon* would have been aggressive towards each other, fighting over food, territory and mating rights. Wear marks on their teeth suggest that they habitually ground them against one another, possibly in order to frighten rivals.

Fossilized *Hyaenodon* skeletons have been found across the world, but especially in the USA and eastern Asia. They were creodonts, a group of primitive placental mammals that evolved around 60 million years ago and went extinct around 38 million years later. During that time the creodonts were the dominant meat-eating mammals on Earth. Aside from giants like *Hyaenodon*, there were many dozens of other creodont species, all of which were equipped with powerful jaws, sharp teeth and claws. The earliest creodonts were small and cat-like, and lived in the jungles, but as time progressed and the forests shrank, so the creodonts grew larger, until they became the top predators. *Hyaenodon* was one of the few creodont species to survive the small extinction event that took place 34 million years ago, at the end of the Eocene epoch. For a while afterwards it was very successful, but it was eventually outcompeted by the faster and more efficient carnivora predators, such as the cats (see **Dinofelis**) and bear-dogs (see **Cynodictis**).

Above > Two *Hyaenodon* in full flight gain ground on their prey, an **Entelodon**; for millions of years *Hyaenodon* was a top predator.
Below > The environment *Hyaenodon* lived in was hot and dry; hunting would have taken place in the evening or early morning.

Fascinating Fact > Wear on fossilized teeth suggest *Hyaenodon* may have ground its teeth together in order to threaten rivals.

Entelodon
The rhino-sized pig

name	Entelodon (en-TELL-oh-don), meaning 'perfect-toothed'		size	2 m (6.5 ft) tall
animal type	Placental (artiodactyl) mammal		diet	Omnivorous
lived	45–25 million years ago		fossil finds	North America and Asia

Entelodon has been described as the 'pig from hell'. It was aggressive, could reach the size of a rhino and could eat almost anything. Its proportions are staggering: it could reach 1 tonne in weight, and the skull alone could be 1 m (3.3 ft) long.

Entelodon was like a moving waste-disposal unit, capable of eating anything from fallen fruit to dead bodies. Its thickly enamelled teeth are often broken, which probably resulted from crunching bones, while other teeth have wear marks that show they also ate nuts, roots and vines. They were fast-moving and could have hunted other animals, crippling them with a bite from their long, robust jaws. The strong neck muscles meant that no part of a kill would be wasted; even the most solid bone could be crushed to a pulp.

Fossilized footprints and other evidence suggest that *Entelodon* lived in small family units, which would have wandered the open plains foraging for food. They were especially fond of watering-holes where thirsty animals would congregate to drink. Here *Entelodon* could have hunted other animals, or have scavenged the remains of other predators' kills. Given the chance, they would even turn cannibal and eat one other.

The skeletons of most *Entelodon* show terrible wounds that could have been caused only by other *Entelodon*. Smashed cheekbones, bite marks and severe head wounds were commonplace injuries for these animals, and were almost certainly the result of violent fights over food or mating rights. In one instance, an *Entelodon* managed to clamp its jaws across the snout of its rival; its teeth created wounds 5 cm (2 in) deep. Fortunately, *Entelodon*'s long legs meant that it could usually escape danger by running away.

Fascinating Fact > Evidence from fossil skulls shows that *Entelodon* would frequently fight each other, inflicting terrible damage.

The fossils of *Entelodon* are sometimes found in abundance, especially in North America, where hundreds have been recovered. *Entelodon* belongs to the artiodactyls, a broad group of even-toed placental mammals that includes pigs, camels, deer and cows. The artiodactyls are related to perissodactyls, or odd-toed mammals (see ***Propalaeotherium***), which include horses, tapirs and rhinos. Together the artiodactyls and perissodactyls form the ungulates or hoofed mammals.

Although it is often described as a pig, *Entelodon* is only a distant relative of the farmyard swine and boars that are around today. It belongs to a family called the entelodonts that, around 45 million years ago, shared a common ancestor with the modern pig family. There were many species of entelodont, most of which were large, brutish-looking animals that fulfilled a role as both hunters and scavengers. Despite their success, the entelodonts were all extinct by 23 million years ago, possibly because they had to compete with their cousins the pigs.

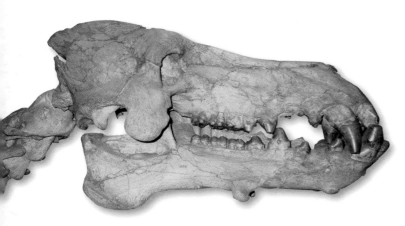

Left > The skull of the average entelodont could contain many traumatic wounds, often caused by encounters with its own kind.
Right > An *Entelodon* and **Hyaenodon** confront one another; given their size and strength it would have been an evenly matched fight.

Indricotherium
The largest ever land mammal

name	Indricotherium (IN-drik-oh-THEER-ee-um), meaning 'Indrik's beast'	size	Males 4.5 m (15 ft) tall; females 4 m (13 ft) tall
animal type	Placental (perissodactyl) mammal	diet	Herbivorous
lived	30–25 million years ago	fossil finds	Mongolia

Indricotherium is the largest-known land mammal of all time (some marine mammals, such as the whales, grew bigger). Male *Indricotherium* stood taller than a house and could weigh up to 15 tonnes, which is about the same as the dinosaur ***Diplodocus***. It was vegetarian, and had long legs and an elongated neck to help it reach up to the top of trees.

The climate in which *Indricotherium* lived was hot and dry, with a short wet season and a long dry one. At certain times of the year, therefore, finding enough food and water would have been tricky. Being so large helped with this, as it meant that *Indricotherium* could

store enough fat, water and other energy reserves to help see it through the lean times. When food was available, it could eat large quantities in a single sitting. Too many *Indricotherium* in one area would soon result in the trees being stripped bare of their leaves, so it is likely that the animals were solitary and roamed long distances looking for fresh feeding grounds.

The size of *Indricotherium* presented some problems. Being warm-blooded meant that its body gave out heat constantly, so it would have been difficult for a large animal to keep itself cool in hot weather. It might therefore have chosen to move about at night, when it was cooler. It is thought that it would be difficult for a land mammal to get any larger than *Indricotherium* without overheating.

Adaptations in its ankle structure meant that *Indricotherium* could walk for long distances and also lie down on the ground to rest and sleep. It probably had a good memory that would allow it to return to watering-holes or favourite feeding grounds.

Male *Indricotherium* had especially thick skulls that they could have used in courtship battles, throwing their heads sideways into the flanks of their rivals, just as modern giraffes do. Based on their size, the females were probably pregnant for around two years, and gave birth to a single calf that they would look after for several years.

Below > Like modern rhinos, *Indricotherium* would probably only have had one calf at a time, which would have been cared for over a number of years by the mother.

Indricotherium fossils are rare, having been found only in Mongolia, but the bones of some smaller related species have been found across Europe and Asia. *Indricotherium* was a perissodactyl (odd-toed) hoofed mammal, and is thus distantly related to the horses (see **Propalaeotherium** for more details), brontotheres (e.g. **Embolotherium**) and chalicotheres (e.g. **Chalicotherium**). *Indricotherium* is a member of the rhinoceros family (although it had no horn), and it is sometimes referred to as a 'running rhino' because of its long legs. For complex reasons *Indricotherium* has recently been renamed *Paraceratherium*, but most scientists still call it by its old name.

Cynodictis
A rabbit-chasing predator

name	Cynodictis (sy-no-DIK-tis), meaning 'in-between dog'
animal type	Placental (carnivora) mammal
lived	28–23 million years ago

size	30 cm (1 ft) tall
diet	Carnivorous
fossil finds	North America and Asia

Cynodictis was a small, carnivorous, dog-like mammal that could run very fast and dig efficiently. It used its speed to chase down rabbits and small rodents, but may also have been able to dig them out of their burrows. *Cynodictis* lived on open, semi-arid plains that were crisscrossed by rivers.

Using its digging skills, *Cynodictis* would make itself dens in steep riverbanks, which it would line with moulted fur and vegetation. In here the female *Cynodictis* would give birth to a litter of around five pups, which she would feed and protect for several months, suckling them at first, then later bringing them food. Unfortunately, the dens would sometimes be destroyed by flash floods that killed all the animals inside, but preserved them as fossils.

Cynodictis was part of a group of placental mammals known as amphicyonids (also called bear-dogs because some species looked like dogs but were the

size of a grizzly bear), which first evolved around 40 million years ago. They were members of the carnivora, the order of mammals that contains swift-moving predators, such as the cats, dogs and bears; they probably evolved from the creodonts, a group of mammals that includes **Hyaenodon**.

The oldest carnivora fossils date from around 55 million years ago. Most of the early species were small, cat-like creatures that hunted on the ground and in trees. Around 40 million years ago, the carnivora diversified, giving us the ancestors to cats, bear-dogs, dogs, bears and mongooses, and also the seals, walruses and sea lions.

The bear-dogs were more closely related to dogs and bears than they were to cats; some bear-dog species grew to 4 m (13 ft) or more in length. Around 9 million years ago (the Late Miocene epoch) the bear-dogs became extinct.

Chalicotherium
A huge knuckle-walking herbivore

name	*Chalicotherium* (KAL-ik-oh-theer-ee-um), meaning 'pebble beast'	size	Male 2.6 m (8.5 ft) tall; female 1.8 m (6 ft) tall
animal type	Placental (perissodactyl) mammal	diet	Herbivorous
lived	25–18 million years ago	fossil finds	Europe and Asia

Chalicotherium was a large, slow-moving mammal that roamed the open plains of Asia and central Europe. It had large hands with curved claws that were twisted inwards at a right angle. Although four-footed, its twisted hands meant that *Chalicotherium* was forced to walk on its knuckles, with its claws bent in towards the wrists. This meant that it carried much of its weight on the hind legs and, although it was a powerful animal, it could not run and was thus vulnerable to attack from large predators, such as **Hyaenodon**.

When eating, *Chalicotherium* would sit on its haunches and use its long arms to pull high-up branches towards its mouth, a bit like gorillas and pandas do. *Chalicotherium* had no front teeth in the upper jaw, and even the back teeth show little sign of any wear and tear. This suggests that they must have been fussy eaters, picking only the newest, freshest shoots and putting them into the back of their mouths. With such a large bulk to maintain, *Chalicotherium* would have spent almost its entire life eating or searching for food.

In one location in Slovakia over 60 *Chalicotherium* skeletons were found preserved in the same geological location. Some people have taken this to mean that they travelled in herds, but it is now thought more likely that these skeletons were the result of individual animals stumbling into the same deep crevice. In life they were probably either solitary or moved together in small groups. Male *Chalicotherium* were much larger than the females, and probably fought each other during the mating season.

The chalicotheres, the group of mammals to which *Chalicotherium* belongs, were perissodactyl mammals, which places them in the same broad group as the horses (e.g. **Propalaeotherium**), rhinos (e.g. **Indricotherium** and **Coelodonta**) and the brontotheres (e.g. **Embolotherium**). The chalicotheres first evolved around 45 million years ago in central Asia. Around 17 million years later they split into two main groups, one of which contained knuckle-walkers, such as *Chalicotherium*. The other group, of which **Ancylotherium** was a member, walked normally on flat feet.

Fascinating Fact > With no front teeth and no wear on its back teeth, it is clear that *Chalicotherium* fed on only the softest new shoots and leaves.

Deinotherium
The largest elephant species

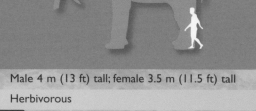

name	*Deinotherium* (dy-noh-THEER-ee-um), meaning 'terrible beast'		size	Male 4 m (13 ft) tall; female 3.5 m (11.5 ft) tall
animal type	Placental (proboscidian) mammal		diet	Herbivorous
lived	20–2 million years ago		fossil finds	Europe and Africa

At 4 m (13 ft) tall and weighing around 10 tonnes, *Deinotherium* is the second largest land mammal of all time. It is a cousin of the modern elephant, and had many elephantine characteristics, including tusks, a thick hide, long legs and a trunk.

Aside from its size, *Deinotherium*'s most noticeable characteristic was two downward-pointing tusks. These were in the lower jaw (unlike modern elephants, whose tusks are in the upper jaw) and had distinctive wear marks on them. They were probably used for stripping the bark from trees, which *Deinotherium* would then eat, as well as for self-defence and sexual display. It was at one time thought that the tusks could have been used to dig up roots and tubers, but *Deinotherium*'s mouth was too high off the ground to make that practicable.

The first *Deinotherium* fossils were discovered in the 1820s in Germany. At the time, the dinosaurs were unknown, and scientists were amazed at the size of their find, so they gave it the name 'terrible beast'. Fossils of *Deinotherium* were later found in Africa, often in the same rocks as hominids, such as *Australopithecus*. Around 10 million years ago the climate began to cool and the range of *Deinotherium* shrank, until it became restricted to a small region in eastern Africa. It went extinct at the start of the ice age, around 1.5 million years ago.

Fascinating Fact > *Deinotherium* was the second-largest known land mammal, the first being *Indricotherium*.

Deinotherium was a proboscidian mammal, which means that it is related to living elephants, as well as to extinct species, such as **Moeritherium** and **Mammuthus** (the woolly mammoth). However, *Deinotherium* is not a direct ancestor of modern elephants: it is actually part of a family known as the deinotheres, a sister group to the elephantids (modern elephants and mammoths). The deinotheres and elephantids share a common ancestor that lived around 30 million years ago.

As a whole, the proboscidians (i.e. deinotheres and all other elephantine species) have a complex evolutionary history. They first evolved around 40 million years ago on the African continent, which at the time was separated from the rest of the world by the Tethys Sea. The proboscidians evolved in isolation in Africa, producing weird and wonderful new species, such as the gomphotheres, whose bottom teeth were large, flat, and stuck out horizontally, a bit like a shovel.

During the Miocene epoch (24–5 million years ago), the Tethys Sea would periodically close up, forming a land-bridge between Africa and Europe. The deinotheres, gomphotheres and elephantids (plus other types of proboscidian) made it across this land-bridge and began to spread themselves about the globe. Some, such as **Mammuthus**, were extremely successful, and even managed to make their way down to South America. Others, such as *Deinotherium*, did not get further than Europe. The Miocene was the proboscidians' most successful time, with dozens of species being found across the world. When the world began to get colder, around 5 million years ago, most proboscidians found it hard to adapt, and many species started to become extinct. Only the elephantids (mammoths and elephants) were able to survive the coming ice age, and of these only two species remain living (the African and Indian elephants).

Left > *Deinotherium*'s downward tusks were most probably used to strip bark from trees, much to the grief of any arboreal animals.

Right > A hormone-enraged *Deinotherium* reacts violently to the nearby presence of a male and female **Australopithecus afarensis**.

Ancylotherium
An African giant herbivore

name		Ancylotherium (AN-sy-loh-THEER-ee-um), meaning 'hooked beast'	size		2 m (6.5 ft) tall
animal type		Placental (perissodactyl) mammal	diet		Herbivorous
lived		6.5–2 million years ago	fossil finds		Africa

Ancylotherium was a large hoofed animal that lived in the grassland regions of tropical Africa. It had long, straight front legs and short back legs, which gave it a pronounced sloping back. It probably lived by browsing on trees and bushes, and would have wandered about the East African plain in small herds, looking for clumps of fresh foliage on which it could feed.

There were few large predators in East Africa that could have tackled an adult *Ancylotherium*, but the juveniles were almost defenceless, and would have been vulnerable to attack by large cats (e.g. **Dinofelis**). For this reason, it seems probable that the young *Ancylotherium* would have sought the protection of their parents or other adults for some time after their birth. Even so, *Ancylotherium* was not likely to have been a herd animal, and probably either lived a solitary life or moved around in small groups.

Ancylotherium was one of the last surviving species of chalicothere, a family of perissodactyl mammals that includes **Chalicotherium**. The chalicotheres may have been outcompeted by the arrival of more efficient herbivores, such as the ancestors of the zebra, gazelle, wildebeest and buffalo that dominate the plains of modern Africa. *Ancylotherium* fossils are very rare and known only from Kenya, Tanzania and Ethiopia. Here they are often found in the same rocks as early hominids, such as **Australopithecus afarensis**, which suggests that our distant ancestors were familiar with these large beasts.

Fascinating Fact > Although nothing like *Ancylotherium* exists today, our ancestors would have been very familiar with it.

Dinofelis
A lethal sabre-tooth carnivore

name	Dinofelis (dy-noh-FEE-liss), meaning 'terrible cat'
animal type	Placental (carnivora) mammal
lived	5–1.4 million years ago

size	70 cm (2.4 ft) tall
diet	Carnivorous
fossil finds	Europe, Asia, North America and Africa

Dinofelis was a medium-sized but powerful cat that possessed two prominent sabre teeth. It lived in forests and open woodlands across the world. It was a lone hunter, and would stalk its prey carefully before pouncing and swiftly killing it with its sharp claws.

The front limbs of *Dinofelis* were particularly strong and muscular, allowing it to deliver debilitating strikes with its claws, and also to pin struggling prey to the ground. Its short sabre-teeth could have been used to help deliver fatal wounds, especially to the neck region.

The favourite prey of *Dinofelis* was large mammals, such as antelope, pigs and juvenile horses. It was not a fast runner, so would have to spend time stalking its victims before making a sudden leap. In eastern and southern Africa *Dinofelis* would also hunt primates, such as baboons, and hominids, such as ***Australopithecus afarensis***, which would have had few defences against a large cat. Once killed, the prey would be taken to a place of safety, such as up a tree or into a cave, where *Dinofelis* could eat in peace. However, it was a messy eater and would scatter bones everywhere, many of which were later chewed on by scavengers such as birds and hyenas.

Dinofelis was very successful, and its fossils have been found around the world. However, it is its association with African hominids that really interests scientists. Several fossil sites from South Africa seem to show that *Dinofelis* was very fond of hunting ***Australopithecus afarensis***. It is thought that the gradual disappearance of the forests in which *Dinofelis* hunted may have contributed to its extinction at the start of the ice age.

Dinofelis was a member of the felid (cat) family, a group of highly successful carnivora mammals, which first evolved around 35 million years ago (the early Oligocene epoch). It is thought that as a group the cats shared a common ancestor with the civets and the hyenas, but during the Oligocene and Miocene epoch they grew more agile and larger, until in many parts of the world they became the top predators (see also ***Smilodon***). Their success as hunters is believed to have led to the decline of many other predatory groups, such as the creodonts (e.g. ***Hyaenodon***) and the bear-dogs (e.g. ***Cynodictis***).

Fascinating Fact > Fossil evidence suggests that *Dinofelis* fed on our ancestors.

Australopithecus afarensis
The original missing link

name		Australopithecus (oss-trah-loh-PITH-ek-us) meaning 'southern primate'	size		Male: 1.5 m (5 ft) tall; female: 1.2 m (4 ft) tall
animal type		Placental (primate) mammal	diet		Omnivorous
lived		3.9–3 million years ago	fossil finds		East and South Africa

Australopithecus afarensis was an early species of hominid, and thus a distant relative of our own species, **Homo sapiens**. Like us, it walked upright on two legs for much of the time – an efficient method of moving about, which also made it taller and thus allowed it more easily to see danger coming. Even though it was bipedal, *Australopithecus afarensis* still spent much of its time living in trees, where it would sleep or, in an emergency, hide.

The diet of *Australopithecus* consisted mostly of soft fruit, nuts and any other food that it could obtain from trees and bushes. However, it would also eat insects and small animals, and even scavenge meat from dead bodies. This could bring dangers with it: one fossil hominid is thought to have died of excess vitamin A poisoning after eating the liver of a carnivore.

The eyesight of *Australopithecus* was particularly acute, and its hand–eye coordination was superb. This not only allowed it to move about in trees at speed, but also meant it could manipulate small objects, such as nuts and fruit, in its hands. Although *Australopithecus*'s brain was of a similar size to that of a chimpanzee, it had yet to learn how to use complex tools.

Australopithecus afarensis was just one of several similar species that lived in Africa between 4.5 and 1.5 million years ago. Among these was *Australopithecus robustus*, known as 'nutcracker man' because its jaws were so heavily set that some used to think it could have cracked nuts with its teeth (which it almost certainly didn't).

One of the best pieces of evidence about *Australopithecus* comes from a set of 3.6-million-year-old fossilized footprints found in Tanzania. These show two adult *Australopithecus* walking side by side, while a smaller one (presumably a child) walked behind, playfully stepping in its parents' footprints. This not only confirms that *Australopithecus* walked upright, but also gives us a rare insight into the lives of these ancient hominids.

Left > The discovery of the first *Australopithecus afarensis* skeleton (nicknamed Lucy) was a scientific revelation. Since then other older hominid fossils have been recovered.

Opposite > *Australopithecus afarensis* could walk upright but still spent much of its time in trees.

Australopithecus is part of the ape group of primates, whose evolutionary history is complex and the subject of much scientific disagreement. In simple terms, the apes split from the 'old world monkeys' (see **Apidium**) around 35 million years ago. Then, at around 20 million years ago, the evolutionary branch that led to the orangutan split off, followed, around 10 million years ago, by the one that produced the gorilla.

Of all the living apes, modern humans are closest to the chimpanzees and share around 99 per cent of their DNA. However, working out when the split between hominids and chimps occurred is difficult. The best current estimate is that it happened between 7 and 9 million years ago. After this, the chimps' evolutionary tree went one way, while that of the hominids produced (among others) *Australopithecus* and, in time, humans such as **Homo neanderthalensis** and **Homo sapiens**. The first *Australopithecus* fossils were discovered in South Africa in 1925, and provided some of the first solid evidence that humans were descended from apes. In 1974 the discovery of a female *Australopithecus afarensis* skeleton in Ethiopia took the human evolutionary tree back to 3.2 million years ago. Currently the oldest-known fossil hominid is *Ardipithecus*, which lived in Africa around 5.8 million years ago.

Fascinating Fact > The most famous fossil of *Australopithecus afarensis* was named Lucy after the Beatles' song 'Lucy in the Sky with Diamonds'.

Carcharodon megalodon
A whale-killing shark

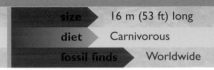

name	*Carcharodon megalodon* (kark-a-row-don meg-a-la-don)		**size**	16 m (53 ft) long
animal type	Neoselachian (lamnid) shark		**diet**	Carnivorous
lived	16–1.6 million years ago		**fossil finds**	Worldwide

Carcharodon megalodon (also called the megalodon shark) is the largest known of all the predatory sharks. It had a streamlined, muscular body and looked like a gigantic version of the much-feared great white shark. Weighing around 48 tonnes, it spent much of its time feeding.

This shark was a powerful predator, capable of making high-speed attack runs. Its mouth could open to over 2 m (6.5 ft) wide, revealing a set of giant teeth, some of which were 21 cm (8.5 in) long and which were sharp enough to cut through the toughest flesh and bone.

The megalodon shark liked to prey on the dolphins and large whales that lived in deep offshore waters. It would cruise around, waiting for one of them to surface. Once a victim was spotted, megalodon would come at it from underneath, swimming vertically upwards at speed. At the last moment the giant jaws would open and the teeth would be rammed into the prey, tearing out large chunks of its flesh. On its initial attack run, megalodon would probably go for vulnerable parts of the body, such as the flippers and tail. This would immediately cripple the animal so that it couldn't swim away or dive downwards. Megalodon's teeth would often fall out during these attacks, but new ones would grow in their place.

Juvenile megalodon were not big enough to attack whales. Their teeth are usually found in shallow waters, suggesting that they lived inshore. Here they probably hunted large fish and smaller marine mammals, such as **Odobenocetops**.

Fossilized teeth from megalodon (which are similar to those of the great white shark) have been found around the world. From their distribution it is possible to see that megalodon preferred tropical and temperate seas, and did not stray into cold waters. It could, however, live many kilometres from the coast in the open ocean. Other than its teeth, few parts of its body are known. Early estimates gave megalodon lengths of up to 25 m (82 ft), but 16 m (53 ft) is now thought to be more realistic.

When whales moved from the tropics to the polar regions around 2 million years ago, megalodon could not follow them because it was a warm-water shark. With its main source of food gone, it soon became extinct.

Megalodon was a neoselachian shark, one of the group of fish that contains all modern sharks and rays. The neoselachians are an ancient group that first evolved in the Carboniferous period (354–290 million years ago), having split away from other shark lineages, such as the hybodonts (e.g. **Hybodus**). The jaws of the neoselachians could open particularly wide, which, combined with their sharp teeth, allowed them to attack and swallow large fish. The neoselachians contain many hundreds of species, which have divided and subdivided into around 35 separate families. The megalodon shark belongs to the lamnidae, the family that contains many living and fossil predatory sharks, including the great white shark (*Carcharodon carcharias*), the largest living carnivorous species.

Left > The soft cartilaginous skeleton of *Carcharodon megalodon* means that only fossils of its teeth are usually found. From these it is easy to reconstruct the size of its jaws, which, like those of many sharks, would have had not one but several rows of teeth. Some fossilized teeth look so fresh it has been suggested that megalodon is still alive, although this is unlikely.

Odobenocetops
Pliocene shark bait

name	*Odobenocetops* (oo-de-oh-ben-oh-sey-tops)
animal type	Placental (cetacean) mammal
lived	5–3 million years ago

size	2.1 m (7 ft) long
diet	Carnivorous
fossil finds	Peru

Odobenocetops was a large marine mammal that cruised the shallow coastal waters off southern Peru. It was a powerful swimmer, but would need to surface periodically to take a breath through its blow-hole. Although it had excellent eyesight, it could also find its way about in dark and murky waters by using the 'melon' organ in its head, which it would have used (just as dolphins do) to navigate the seabed using echolocation.

It is the giant tusks of *Odobenocetops* that make it so interesting to scientists. On males the right-hand tusk could reach 1.2 m (4 ft) long, while the left-hand tusk was only 25 cm (10 in) long. The tusks were quite fragile, so may have been for display rather than defence, or for use in mating battles. The females had only two small tusks.

Odobenocetops foraged on the seabed, where its flat belly would permit the mouth and fat lips to grub in the mud for clams and other shellfish. Once found, each would be held in the lips and the prey sucked from its shell.

The first *Odobenocetops* fossils were found in 1993, but the animal is known only from several skull specimens and a handful of bones from its body. It may have been the prey of the **Carcharodon megalodon**, whose fossilized teeth have been found in the same rocks.

Odobenocetops was a type of dolphin, and thus belongs to the odontoceti (toothed whales), which are themselves part of the cetacean mammals. The dolphins and porpoises split away from the other toothed whales around 20 million years ago. Being smaller and quicker than their larger whale cousins, the dolphins were able to hunt shoals of fast-moving fish in shallow waters. Most dolphins and porpoises, be they fossil or living, have the same approximate body shape, which makes an evolutionary offshoot such as the slow-moving and clam-eating *Odobenocetops* a rarity.

> **Fascinating Fact** > The males used their long tusks to joust with each other during the mating season.

Top right > *Odobenocetops*'s mouth was underneath its head; it would feed by grubbing around on the sea floor with its mouth.
Right > *Odobenocetops* would have made an ideal meal for the giant shark **Carcharodon megalodon**.

Smilodon
The biggest of the sabre-toothed cats

name	Smilodon (SMY-loh-don), meaning 'knife tooth'		size	1.2 m (4 ft) tall
animal type	Placental (carnivora) mammal		diet	Carnivorous
lived	2.5 million–100,000 years ago		fossil finds	North and South America

Smilodon was a large, predatory cat that lived on the open grassy plains of the eastern USA and South America. It was a voracious predator that attacked large mammals, such as deer and antelope. It probably lived in prides, much as modern African lions do. There is evidence that older or injured cats were supported by other members of the pride.

When hunting, several *Smilodon* would have worked in cooperation with each other, stalking their prey through the long grass, then taking it in turns to give chase. When their prey was exhausted the *Smilodon* would move in and use their 21-cm (8.5-in) sabre-teeth to kill it. However, as these teeth were easily broken, they did not use them to rip or stab at a struggling body. Instead, they probably wrestled the prey to the ground with their powerful forelimbs, then clamped the sabres around their prey's windpipe. They would then use their powerful jaw muscles to strangle it to death or rip out its throat.

Having such long teeth meant that *Smilodon* could not risk biting on to bone, as this might actually snap a tooth (some fossils do have broken teeth). Instead, it ate meat only from the softest parts of the body, such as the stomach region. This would leave plenty of food on the carcass for other scavengers, such as *Phorusrhacos*.

Left > The sabre teeth of *Smilodon* were fragile and could only be used on the soft part of their prey's body.

Above > Here a *Smilodon* is in hot pursuit of a **Macrauchenia**, which is attempting to escape by making sudden changes in direction. *Smilodon* could only run at speed in short bursts; if the *Macrauchenia* has enough stamina it should evade capture.

Hundreds of *Smilodon* fossils have been found across the USA and South America, but the most famous ones come from the La Brea tar-pit in California. Thousands of years ago La Brea was a sticky swamp in which large animals would sometimes become mired. These trapped animals would attract predators, such as *Smilodon*, which would enter the swamp and then get stuck themselves. Many hundreds of animals could become bogged down this way.

There are so many *Smilodon* fossils in La Brea that scientists have been able to deduce much about its lifestyle. For example, some specimens had such serious injuries that the animals would not have been able to hunt effectively, yet they obviously managed to stay alive for some years after being injured. This probably means that they were feeding from kills made by other *Smilodon*, which in turn suggests that they were living in prides.

Although sometimes called the 'sabre-toothed tiger', *Smilodon* is not in fact closely related to living tigers at all. While it is part of the felid (cat) family, it is only distantly related to living cats, such as lions and tigers. At one time there were several species of *Smilodon* living in eastern North America. When South America was joined to North America, around 2.5 million years ago, at least one species of *Smilodon* made the journey south and settled in the grassland regions of Argentina and Brazil. Its fossils are relatively common, and *Smilodon* was evidently a successful predator, hunting large animals such as **Macrauchenia**. The reason for its extinction remains a mystery, although much of the large prey on which it fed was also dying out. *Smilodon* is sometimes shown as having attacked cavemen, but this was impossible as it went extinct long before the first humans arrived in America.

Phorusrhacos
A dinosaur-like carnivorous bird

name	Phorusrhacos (FOR-uss-RAH-kuss), meaning 'rag-bearer'	size	3 m (10 ft) tall
animal type	Neognath (phorusrhacid) bird	diet	Carnivorous
lived	5 million–400,000 years (possibly 15,000 years) ago	fossil finds	South and North America

Phorusrhacos was a gigantic meat-eating bird that stood over 3 m (10 ft) tall and had a sharp, hooked beak. It was flightless and lightly built, with long legs and tightly fitting feathers, all of which helped it to run fast. For millions of years *Phorusrhacos* was the top predator on the South American grassy plains.

As a close relative of the living secretary bird, *Phorusrhacos* is commonly thought to have used the same violent hunting technique, which involves sneaking up on prey through long grass. When in range of its victims, which included small horses and **Macrauchenia**, *Phorusrhacos* would launch an attack, running at speeds of up to 70 kph (43 mph) and using its beak to bite and slash, thus disabling the animal it was chasing. Once trapped, the prey would be grabbed in the powerful beak and shaken violently before being eaten.

One unusual aspect of *Phorusrhacos*'s biology is that it had a sharp, stout claw on its wing. This would have been of little use for hunting or defence, so it is thought to have played a part in its courtship ritual.

The first fossils of *Phorusrhacos* were found in Argentina in 1887, but the best-preserved specimens come from a flooded cave, in Florida, which can be reached only by scuba divers. The oldest phorusrhacid birds (the family to which *Phorusrhacos* belongs) are 50 million years old, and are found in Europe.

> **Fascinating Fact** > The first scientists to study *Phorusrhacos* gave it the nickname 'terror bird'.

The range of the phorusrhacids then decreased, so from around 27 million years ago they were found only in South America, which was then cut off from the rest of the world for nearly 30 million years.

It was only 2.5 million years ago that North and South America collided. After this event *Phorusrhacos* moved north into the southern United States (where it is sometimes referred to under its redundant name of *Titanis*), and thrived there until 400,000 years ago. (A recent find suggests that it may have been alive as recently as 15,000 years ago.)

Phorusrhacos is just one of a number of large animals that went extinct during the last ice age. Other famous casualties include the mammoths (e.g. **Mammuthus**), woolly rhinos (e.g. **Coelodonta**), giant sloths (e.g. **Megatherium**), sabre-toothed cats (e.g. **Smilodon**) and glyptodonts (e.g. **Doedicurus**). In fact, so many animals have gone extinct that some scientists think that this could be part of a new mass extinction event. However, the cause of these extinctions is not certain.

It used to be thought that humans might have helped hunt some of these animals into extinction, but in fact many died out in areas where no humans were then living. The pattern of these extinctions remains puzzling. Some large predators, such as *Phorusrhacos* and **Smilodon**, might have become extinct because they could not compete with small, swifter hunters, such as members of the cat and dog families. Some of the ice-age giant herbivores, such as **Megatherium**, mammoths and woolly rhinos, probably became extinct because the specialized grasslands on which they fed disappeared around 10,000 years ago. Whatever the real cause of the extinctions, the ice age appears to have had a much worse effect on larger animals than on smaller ones. This is a characteristic of many mass extinction events, including the one that killed the dinosaurs 65 million years ago. It would seem that when there are upheavals in the world's ecosystems, it is the biggest animals that are the first to go extinct.

Left > A pair of *Phorusrhacos* fight over the carcass of a **Smilodon** cub. They would not have dared to attack an adult.
Right > The cat **Smilodon** and bird *Phorusrhacos* were top predators.

Megatherium
The famous giant sloth

name	Megatherium (meg-ah-THEER-ee-um), meaning 'giant beast'		**size**	6 m (19.5 ft) long
animal type	Placental (xenarthran) mammal		**diet**	Herbivorous
lived	1.9 million–8000 years ago		**fossil finds**	North and South America

Megatherium is a relative of the present-day tree sloths of South America, but, at around 6 m (19.5 ft) long and 4 tonnes in weight, was considerably bigger. It spent its life wandering the open forests and plains of South America and the southern United States.

Its diet was predominantly vegetarian, and consisted of leaves, flowers and young branches. To reach them *Megatherium* would rise on to its hind legs and use its 70-cm (2.4-ft) claws to pull down high branches towards its long, grasping tongue. The fossilized dung from one animal indicated that it had eaten over 70 different types of plant. It is also possible that in times of drought *Megatherium* could have scavenged flesh from dead bodies.

Giant ground sloths were the largest land animals in their environment, but they nevertheless had some formidable defence strategies. Their giant claws were attached to powerful arms and could be used to stab

or swipe at predators. Underneath their long, shaggy hair was skin impregnated with thousands of small bony plates, which served as chain mail against attacking teeth or claws. As a final show of aggression, *Megatherium* could rear up and walk on two legs for short periods of time.

Even so, this giant sloth could not defend itself against spears and clubs, and it may have been hunted to extinction by the first humans to arrive in South America. However, given the changes that were taking place in the world's climate at this time, it seems more probable that *Megatherium* was pushed into extinction by changes in the ecosystem in which it lived.

Fascinating Fact > *Megatherium* had small bony lumps under its skin that acted like natural chain mail.

The first *Megatherium* skeleton was discovered in Argentina in 1787; before then people had no idea that extinct animals could grow so large. When the first *Megatherium* skeletons were exhibited in European museums they caused a sensation – just as well when the purchase of one skeleton could consume a museum's entire annual budget. In addition to hundreds of fossilized bones, the sloth's skin, hair and dung have also been found preserved in caves, and its footprints preserved on muddy banks. These have allowed scientists to reconstruct the appearance, diet and behaviour of the animal in minute detail.

Megatherium was a xenarthran mammal, which belongs to the group of placental mammals that also contains the armadillos, anteaters and glyptodonts (see **Doedicurus**). The oldest xenarthran fossils date from around 55 million years ago, but the oldest sloth specimens found so far are about 25 million years old. Around 15 million years ago the sloth lineage split, with one branch becoming small tree-dwellers, the ancestors of the modern three- and five-toed tree sloths. The other branch of sloths became ground-dwellers, and soon reached a gigantic size. At one time there were several species of giant ground sloth in existence, but their numbers dwindled. *Megatherium* was the last of the giant ground sloths, but it did survive long enough to overlap with the first humans to reach North and South America. There are myths that *Megatherium* could still survive in remote parts of the Andes foothills, but this seems unlikely.

Top right > Although herbivorous, an adult *Megatherium* was a seriously well-defended animal, which would have had no natural enemies, except perhaps other *Megatherium*. The ground sloths were just one of several species of giant herbivorous mammal living on the Pliocene South American grasslands. Quite how these grasslands were able to support so many large herbivores remains a mystery.

Right > The Victorian public thrilled to the sight of *Megatherium* skeletons like this, and museums would go to great lengths to get hold of them. It was only in the early 1900s, when the first dinosaur skeletons were mounted, that *Megatherium*'s popularity waned.

Bottom right > This trail of fossilized footprints proves the giant sloth *Megatherium* could walk upright on two legs, supporting its weight on the sides of its feet.

Macrauchenia
Sabre-tooth cat food

name	*Macrauchenia* (mak-raw-KEE-nee-ah), meaning 'long llama'
animal type	Placental (litoptern) mammal
lived	7 million–20,000 years ago

size	2.1 m (7 ft) tall (head height)
diet	Herbivorous
fossil finds	South America

Macrauchenia was a four-legged herbivore that was a common sight on the South American plains. Its long neck allowed it to reach high into bushes and trees so that it could browse on leaves. It lived in vast herds that would migrate across the landscape following the seasonal rains.

As it was largely defenceless, *Macrauchenia* was regularly hunted by other predators, such as the swift sabre-toothed cat **Smilodon** and the giant bird **Phorusrhacos**. If danger threatened, its long legs allowed it to run at high speeds, and it had a useful trick that could help it throw off any pursuers. The bones in its legs were arranged so that even when travelling at high speeds it could make sudden swerves and turns, throwing any predators running behind off balance. Even so, *Macrauchenia* that were old, injured or very young must have regularly fallen victim to predators.

The first *Macrauchenia* fossils were discovered by the naturalist Charles Darwin, who was only in his early twenties at the time. At first he thought that they were an extinct type of llama, but he had the bones posted back to London, where they were identified as belonging to a completely different type of animal altogether. In fact, *Macrauchenia* was the last surviving species of a very ancient group of hoofed animals called the litopterns. These were primitive placental mammals that, around 60 million years ago, probably shared a distant common ancestor with the artiodactyl (e.g. **Entelodon**) and perissodactyl (e.g. **Propaelotherium**) mammals. It is thought that the litopterns became stranded in South America when the continent broke away from the rest of the world around 35 million years ago. They remained there for millions of years, but went into decline about 5 million years ago, and by around 20,000 years ago (a few thousand years before the first humans arrived in South America) they became extinct altogether.

Fascinating Fact > Charles Darwin himself stumbled across the first *Macrauchenia* fossils while out riding on the South American plains.

Doedicurus
An armadillo the size of a car

name	Doedicurus (dee-dik-YOO-russ), meaning 'pestle tail'	size	4 m (13 ft) long
animal type	Placental (xenarthran) mammal	diet	Herbivorous
lived	2 million – 15,000 years ago	fossil finds	South America

Doedicurus was a giant armadillo that was the same size and shape as a Volkswagen Beetle, but, at 1.5 tonnes, was considerably heavier. Its shell was made from over 1800 bony plates, each 2.5 cm (1 in) thick, that acted like chain mail. Beneath this armour was a deep layer of fat that acted like padding and could absorb the energy from massive impacts. This made Doedicurus practically invulnerable to attack, but it was also heavy and slow-moving. It would sometimes get stuck on marshy ground or wedged in tight crevices, leaving it vulnerable to starvation or attack. One juvenile Doedicurus that had become stuck in mud was attacked by the sabre-toothed cat **Smilodon**.

As well as armour, Doedicurus had a spiked club tail that could weigh as much as an adult man. All this was not just for protection; during the mating season, males would attack each other with their club tails, inflicting serious damage on each other's shells

(although the bones beneath were rarely hurt). The victors would get to mate with the females. Doedicurus lived only on grass, leaves and perhaps roots. It could not reach high into the trees, so was forced to compete with other browsers for low-level vegetation.

Doedicurus was a glyptodont, a group of xenarthran mammals (see **Megatherium**) that are closely related to living armadillos. The glyptodonts evolved in South America around 20 million years ago, but around 2.5 million years ago they spread into the southern United States. The last of the glyptodonts may have been alive in South America when the first human tribes arrived.

Fascinating Fact > The club-like tail of Doedicurus was a formidable weapon, weighing a staggering 74 kg (163 lb).

195

Megaloceros
Owner of the world's largest antlers

name	*Megaloceros* (meg-ah-LOSS-er-oss), meaning 'giant antler'
animal type	Placental (artiodactyl) mammal
lived	400,000–9500 years ago

size	2.1 m (7 ft) tall
diet	Herbivorous
fossil finds	Europe

Megaloceros was a deer, with the largest antlers of any known hoofed animal. It was one of a number of ice-age mammals that lived on the grasslands and open forest that existed on the ice-sheets' perimeter. It enjoyed short, fertile summers, when food was plentiful, and then endured long, tough winters, when food was scarce.

Fossils of *Megaloceros* are commonly found in Irish peat bogs, which has earned the animal the name of Irish elk, even though it is actually a species of deer.

The antlers, which could reach 3 m (10 ft) across, are found only on males, and were probably used in mating contests or for sexual display. The rutting season would result in males being injured and then being unable to feed themselves; their fossils reveal that they died of starvation. *Megaloceros* is featured in the cave paintings of **Homo sapiens** and was probably hunted by them.

As a deer *Megaloceros* was also an artiodactyl (odd-toed) hoofed mammal (see **Entelodon**). Artiodactyl fossils

are relatively rare until around 15 million years ago, when the first grasslands occur. Grass is nutritionally poor and tough, but the artiodactyls evolved a complex digestive method called rumination, in which the grass would be swallowed, partially digested, regurgitated, chewed again and then passed through a series of stomachs so that the nutrients could be extracted. Being able to ruminate grass gave the artiodactyls a near-inexhaustible food supply. Their numbers expanded hugely, with some species forming themselves into vast migratory herds. It is from these ancient grass-eating artiodactyls that our modern cows, ox, sheep and goats are descended, although years of domestic breeding have removed many of their ancestral traits.

Fascinating Fact > Although known as the Irish elk, *Megaloceros* is not an elk and is not found just in Ireland.

Panthera leo
The cave lion

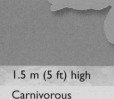

name	Panthera leo (PANTH-er-ah lee-oh), meaning 'panther'		size	1.5 m (5 ft) high
animal type	Placental (carnivora) mammal		diet	Carnivorous
lived	900,000–10,000 years ago		fossil finds	Europe

Panthera leo is the scientific name for the modern African lion of the sort that can be seen on safari today. Today we think of the lion as being an exclusively African animal, but in the past its range extended outside Africa and into Europe. Here it is usually referred to as being the 'cave lion' because its fossils are often found in caves or under rock overhangs.

The cave lion, while being anatomically identical to the African lion, was actually quite a bit larger, and during the winter would almost certainly have had white or near-white fur to help camouflage it against the snow. It would have spent much of its time on the grassy plains and in the woodlands of central Europe, where it would hunt prey, such as deer, horses and even humans. Cave paintings made by **Homo sapiens** show that the cave lions didn't have a bushy mane or tail.

It was not just cave lions that made it to Europe from Africa. During the warmer phases between ice ages, Europe was home to many large African mammals. Archaeological digs in London's Trafalgar Square have revealed that around 400,000 years ago (when the climate was marginally warmer than now) this area was a grassy wetland that was home to lions, hippopotamuses, rhinoceroses, elephants, bison,

wolves and deer. Many of these animals had migrated north from Africa with the warm weather, returning south when the climate cooled again. At that time the Sahara desert was a lush grassland, which meant that many of these animals could live in North Africa, on the border of Europe. Then, around 7000 years ago, the grasslands disappeared and the desert started to form, creating a barrier across North Africa, south of which the lions, hippos and others were trapped. Ancient rock art in the middle of the Sahara desert, made by humans around 8000 years ago, shows a lush landscape full of the sort of large animals that can be seen in game reserves today.

Panthera leo is a member of the cat family, and thus part of the broader carnivora group of mammals, which also includes the bears, dogs and mongooses (see **Dinofelis**). All the modern cats (such as lions, cheetahs, tigers and cougars) are quite closely related to each other, but are only distantly related to the extinct lineages of sabre-toothed cats, such as **Smilodon** and **Dinofelis.**

Fascinating Fact > *Panthera leo* was a very large lion species that would have found humans easy prey.

Homo neanderthalensis
The original caveman

name	*Homo neanderthalensis* (ho-mo ni-AN-der-taal-en-sis)
animal type	Placental (primate) mammal
lived	300,000–28,000 years ago

size	Male 1.7 m (5.5 ft) tall; female 1.6 m (5.2 ft) tall
diet	Carnivorous
fossil finds	Middle East and Europe

The Neanderthals were cave-dwelling humans that lived in ice-age Europe and the Middle East. They were short, stocky and very muscular, which helped them survive against the cold, although they probably also wore animal skins for additional warmth.

The face of a Neanderthal had a big nose, flattened forehead and thick eyebrow ridges. To our eyes these features make the Neanderthals look brutish and stupid. This led to them being portrayed as primitive savages. In fact, archaeological evidence from their campsites shows that they lived in small but socially complex groups, were capable of making fire, spears and flint tools, and looked after sick and injured members of their group.

That said, the stone tools and other implements used by Neanderthals were primitive in comparison to those of **Homo sapiens**. Also, there is no evidence of Neanderthal art, or that they were capable of highly creative thought. Their campsites were usually made in caves or under rock overhangs. They were messy affairs, with food, bones, flints, faeces and other debris being randomly scattered across the floor.

Neanderthals were accomplished hunters, and lived almost exclusively on meat: they ate very few vegetables. They hunted with spears and ate mostly deer and rabbits. On the British Channel Island of Jersey there is evidence that some Neanderthals deliberately drove woolly mammoths (**Mammuthus**) and rhino (**Coelodonta**) off the edge of a cliff to kill them. Hunting could be a dangerous affair, and most Neanderthal skeletons are covered in dozens of old wounds, especially broken bones. The Neanderthals sometimes buried their dead, adding flowers and red minerals to the grave. Often the flesh would be stripped from the corpse, but whether this was for ritualistic or cannibalistic reasons is not known.

The first Neanderthal bones were discovered in Germany in 1856, but they were initially mistaken for those of a dead soldier. Since then, their tools and remains have been found across many parts of Europe, and in some sites in Israel and the Near East.

The evolutionary relationship of the Neanderthals to the rest of the hominids is complex. The seat of hominid evolution is in East Africa, where, around 2.5 million years ago, the first recognizable human (called *Homo habilis*) evolved from **Australopithecus afarensis** or a similar species. *Homo habilis* was still a very primitive

Fascinating Fact > Neanderthals led rough lives, and on average had the same number of broken bones as a rodeo rider.

human, but it was a tool user, something that sets it apart from **Australopithecus afarensis**. Around 1.9 million years ago another hominid species, *Homo erectus*, left Africa and began to spread around the globe, reaching places such as China, Europe and Indonesia. *Homo erectus* (sometimes called 'Peking Man') could use fire. The Neanderthals are descendants of *Homo erectus*, albeit via a complex hierarchy of other human species.

The relationship between modern humans and Neanderthals is controversial. For a brief period of time both Neanderthals and modern humans lived in the same region of Europe. It has often been speculated that modern humans might have deliberately driven the Neanderthals to extinction, or that the Neanderthals and moderns interbred with one another (DNA recovered from Neanderthal fossils currently rules out this last theory). Modern archaeological evidence suggests that the Neanderthals and modern humans had little to do with each other, although just before their extinction the Neanderthals may have picked up on some of the tool-making techniques used by **Homo sapiens**. The latest research suggests that, along with many other large animals, the Neanderthals were unable to adapt to the rapid climatic changes of the ice age, so their numbers gradually declined to nothing.

Right > Neanderthal fossils are commonly found at cave sites.

Mammuthus
The giant hairy elephant of the tundra

name	*Mammuthus* (MAM-oo-thuss), meaning 'earth mole'	size	Male 3.5 m (11.5 ft) tall; female 3 m (10 ft) tall
animal type	Placental (proboscid) mammal	diet	Herbivorous
lived	2 million–6000 years ago	fossil finds	North America, Europe and Asia

Mammuthus, also called the woolly mammoth, was one of the largest land mammals of all time. Its elephantine shape, corkscrew tusks and long, shaggy hair are now symbolic of the ice age.

Mammoths were once a common sight, and moved in small herds across the plains and woodlands of Europe, Siberia and North America, feeding on wild grass, flowers and other low vegetation. There is evidence of a complex social structure within the herds. In the USA the skeleton of one dead mammoth was surrounded by many footprints, indicating that other members of the herd had stood guard over the body (living elephants do something similar).

Mammoths are closely related to present-day elephants, and share many of their features, including the body shape, large size and long, grasping trunk.

However, their ears were very much smaller, and their tusks considerably longer and corkscrew shaped. Their bodies were covered in a dense layer of coarse hair up to 1 m (3.3 ft) long, with an insulating layer of fat underneath the skin. These adaptations allowed mammoths to withstand the cold of ice-age winters, with the tusks probably being used to clear a path through the snow to food underneath.

Being so large, mammoths had to spend most of their time eating. One frozen mammoth was found with 290 kg (638 lb) of grass in its stomach. After the onset of winter, many mammoths would be driven southwards to find new feeding pastures and escape the worst of the cold. It is possible that some mammoths lived to 60 or 70 years old, which is a similar lifespan to that of modern elephants.

The first fossilized mammoth bones were discovered nearly 500 years ago, but they were not identified as belonging to an extinct type of elephant until 1796. Since then, literally thousands of mammoth bones and tusks have been found across the northern hemisphere, and even as far afield as South America. Most famous are the discoveries of frozen mammoth carcasses in Siberia, which occasionally erode out of riverbanks. These frozen time-capsules have been carefully studied by scientists and are now thought to be the remains of animals that were unfortunate enough to fall into icy marshes or crevices and then freeze to death.

Several species of human (including ***Homo neanderthalensis*** and ***Homo sapiens***) either ate or hunted the mammoths; mammoth bones are a common find in their campsites. Humans are sometimes blamed for the decline and extinction of the mammoths, but it is more likely that rapid changes in the environment at the end of the last ice age brought about their demise. A population of pigmy mammoths is thought to have survived on an island off the east coast of Russia until about 6000 years ago.

After the extinction of the mammoth, the only surviving members of the once-numerous proboscidians (the group that contains all the elephants) are the African and Indian elephants. However, the elephants do have some surprising cousins left on Earth. One of these is the marine-dwelling sea cow (or manatee), which split off from the proboscidian lineage around 40 million years ago. An even more unexpected relation is the hyrax, a small, rat-like creature that lives in Africa and split off from the proboscidians around 45 million years ago. Recent DNA tests have proved that both these animals are related to the modern elephant.

Below > Mammoths were social animals that would have lived and travelled in small groups. They preferred to live south of the ice-sheets, in regions that had only a few weeks' snow a year.

Coelodonta
The woolly rhino

name	Coelodonta (see-loh-DON-tah), meaning 'hollow teeth'	size	2.2 m (7 ft) tall
animal type	Placental (perissodactyl) mammal	diet	Herbivorous
lived	500,000–10,000 years ago	fossil finds	Europe and northern Asia

Coelodonta was a large species of woolly rhinoceros that had adapted to living in the ice-age conditions of Europe and Siberia. It was larger and stockier than any modern rhino, had a thick, hairy coat and a very large horn. It ate mostly grass, so needed to live either on plains or in open woodland, where snow would lie on the ground for only a few weeks of the year.

Its most distinctive feature is the large horn which could reach 2 m (6.5 ft) long. It is thought that this may have been used to sweep the ground when it was covered in snow, thus clearing enough space for *Coelodonta* to eat. However, it may also have been used as a weapon. One cave painting shows two *Coelodonta* battling each other, which suggests that, like modern rhinos, they may have been territorial. In the summer they may have shed their shaggy coat in favour of a shorter, cooler one.

The eyesight of *Coelodonta* was probably poor, but its sense of smell was excellent and would have alerted it to any danger, such as approaching **Homo neanderthalensis** and **Homo sapiens**, both of whom are known to have eaten (and therefore probably hunted) this rhinoceros. Other than humans, adult *Coelodonta* were so large that they had little to fear from any predators.

Below > Neanderthals are thought to have hunted *Coelodonta*.

Coelodonta belongs to the perissodactyl mammals, a group that also contains horses, tapirs, brontotheres (e.g. **Embolotherium**) and chalicotheres (e.g. **Chalicotherium**), and is closely related to living rhinoceroses, as well as to extinct ones, such as **Indricotherium**.

The first *Coelodonta* fossils were found in Siberia in 1769, but its long horns were at first thought to be the claws of giant birds. Since then, hundreds of specimens have been found across Europe and northern Russia. One was even found underneath Battersea Power Station in central London. From preserved skin and hair found in Siberia scientists have made some deductions about what *Coelodonta* would have looked like when alive. Other information about its behaviour and appearance has been gained from looking at ancient cave-paintings. The reason for *Coelodonta*'s extinction is not known, although, like the woolly mammoth (**Mammuthus**), it is probably connected to its inability to adapt to the rapid changes in habitat at the end of the last ice age.

Fascinating Fact > *Coelodonta*'s horn could be up to 2 m (6.5 ft) long, but was flat, like a plank of wood.

Homo sapiens
Modern human beings

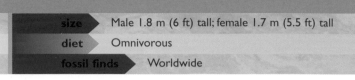

name	*Homo sapiens* (ho-mo SAP-ee-ens), meaning 'knowing man'	size	Male 1.8 m (6 ft) tall; female 1.7 m (5.5 ft) tall
animal type	Placental (primate) mammal	diet	Omnivorous
lived	Around 190,000 years ago – present day	fossil finds	Worldwide

Homo sapiens, our own species, first evolved in eastern Africa around 190,000 years ago. Our ancestors were fantastic toolmakers and very mobile. By 90,000 years ago they had left Africa, and by 40,000 years ago had spread across Asia, into Europe and the Far East.

It is the *Homo sapiens* who made it to Europe that have attracted the most attention because of their complex culture and artistic ability. They are commonly called Cro-Magnon men, after the place in France where their fossils were first identified.

The European Cro-Magnons looked similar to other ancient human species (such as ***Homo neanderthalensis***), but their advantage came not so much from their bodies as from their brains. The Cro-Magnons' ability to think creatively and solve problems resulted in them developing new tools and survival techniques, including advanced flint knives and spearheads, ropes and woven clothes. With these, Cro-Magnons could hunt more efficiently and make more of their food by processing it better. They would even store it for future use. They also supplemented their meat diet with edible plants, fruit and vegetables, which meant that they were less reliant on the dangerous activity of hunting.

The best insight into the brainpower and culture of Cro-Magnon men comes from their cave art and the sculptures and carvings they left behind. The first cave-paintings were discovered in France, and show drawings of animals, such as woolly mammoths (***Mammuthus***) and woolly rhino (***Coelodonta***), as well as more abstract art, such as coloured dots and handprints. Some of these paintings are thought to have had a religious significance, and as such are evidence that Cro-Magnon men were capable of complex speech and abstract thought.

Fascinating Fact > Ironically, *Homo sapiens* is probably the most dangerous predator ever to have evolved.

When ***Homo neanderthalensis*** became extinct around 28,000 years ago, *Homo sapiens* became the principal human species on Earth. (The only other species was Indonesia's ***Homo floresiensis***, whose existence was not discovered until 2003.) Their resourcefulness and use of technology allowed them to survive through the ice age, the end of which (around 10,000 years ago) marks the start of *Homo sapiens'* rise to domination.

During the warmer phases of the last ice age, humans moved themselves across the globe, first settling in Asia and Australia, then, around 11,000 years ago, in North and South America. By 8000 years ago some humans in the Middle East and Europe gave up hunting in favour of a more settled life. They began to farm crops and rear animals, taking away the need for lengthy hunting trips and seasonal migrations. The establishment of farming communities led to a more stable and socially cohesive lifestyle. Populations grew, and by 5000 years ago the first cities appeared in the Middle East. Technology also advanced, going from simple stone, wood and bone tools to complex objects made from metals such as bronze and iron. The first empires began to spring up in the Middle East, North Africa, Asia and southern Europe. As new parts of the world were conquered and colonized, so new technology and cultural lifestyles spread into areas that had hitherto contained small tribal populations. This is a process that is still happening today.

The last 200 years have seen the once disparate human populations of the world all come into contact with each other. Scientific ideas and technologies formulated in the Western world have now been spread globally, but there is still much inequality, with technology-poor countries finding it difficult to compete with their more developed neighbours. The development of technology and science by humans has led to a ballooning of the global population, which currently stands at around 6 billion. This has put pressure on the Earth's limited resources and many of its ecosystems.

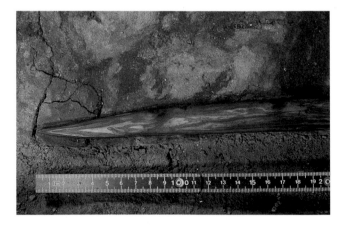

Left > This 400,000-year-old spear, found embedded in horse bones in Germany, is clear evidence that ancient humans could fashion hunting weapons.

Homo floresiensis
The 'hobbit' of Indonesia

name	Homo floresiensis (ho-mo FLOOR-ez-ee-en-sis), 'man from Flores'	size	I m (3.3 ft) tall
animal type	Placental (primate) mammal	diet	Omnivorous
lived	Around 94,000–13,000 years ago	fossil finds	Indonesia

In 2003 a dramatic discovery was made on the remote Indonesian island of Flores. An archaeological team working in a cave there discovered fossils of a previously unknown species of human, which in 2004 was given the name *Homo floresiensis*. This new human looked much like us, but was very much smaller (about half our height) and had a remarkably small brain. Our brain has a capacity of around 1400 cubic cm, while that of *Homo floresiensis* was just 380 cubic cm, giving it the same brain/body ratio as the 4-million-year-old African hominid *Australopithecus afarensis*.

The discovery of a new species of human is not all that unusual. What was surprising about *Homo floresiensis* is that the youngest fossils were only 13,000 years old, which means that these small humans would have lived at the same time and in the same area as our own species. Until this discovery, it was assumed that the last time our species shared the world with another race of humans was during the ice age, when *Homo neanderthalensis* and *Homo sapiens* lived in Europe together.

Little has thus far been deduced about how *Homo floresiensis* might have lived, but it is suspected that they made and used delicate stone tools, many of which were found in the same cave as the fossils. Other aspects of their culture have yet to be properly determined.

The evolutionary relationship between ourselves and this newly found species is as yet unknown. It has been suggested that *Homo floresiensis* might be a distant descendant of *Homo erectus*, a fossil human that lived in the region (and, indeed, across much of Asia) around a million years ago. *Homo erectus* was much taller and had a larger brain, but it is possible that being isolated on an island caused *Homo floresiensis* to become smaller. (Many other animals on Flores have evolved to become either larger or smaller with time; for example, a species of dwarf elephant and a giant monitor lizard have been found there.)

There is even the possibility that *Homo floresiensis* might have been alive until very recently. The tribes on Flores have long spoken about the Ebu Gogo, a race of small, hairy and pot-bellied humans who live in the forests and speak their own unintelligible language. The tribesmen claim that the Ebu Gogo were certainly still to be seen 300 years ago, when the first Dutch settlers arrived, but that there have been sightings in the nineteenth century. Could our own species, *Homo sapiens*, have been sharing the planet with another completely unknown species of human for all this time?

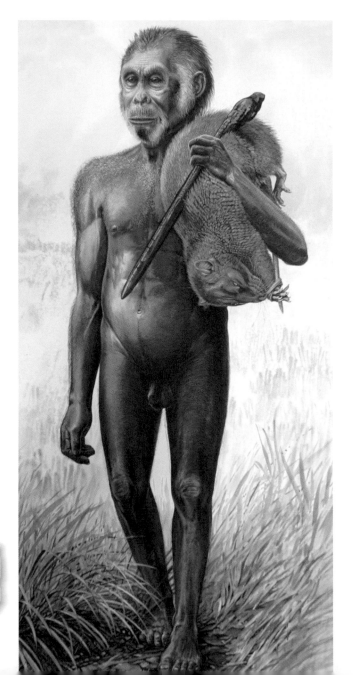

Fascinating Fact > *Homo floresiensis* has been nicknamed 'the hobbit' after the fictional characters created by J.R.R. Tolkien.

Timescale of the Earth

ERA	PERIOD	EPOCH	BEGINNING	MAJOR EVENTS
Cenozoic	Quaternary	Holocene	10,000	HUMAN CIVILIZATION BEGINS
		Pleistocene	1,800,000	ICE AGE; NEANDERTHAL AND CRO-MAGNON MAN EVOLVE; MAMMOTHS ROAM FREE
	Tertiary	Pliocene	5,000,000	OLDEST HOMINID FOSSILS; SOUTH AMERICA JOINS WITH NORTH AMERICA
		Miocene	24,000,000	GRASSLANDS EVOLVE; HOOFED ANIMALS PROLIFERATE; MOUNTAIN CHAINS FORM
		Oligocene	34,000,000	CARNIVORES AND GIANT MAMMALS EVOLVE; ANTARCTIC ICE-SHEET FORMS
		Eocene	55,000,000	FIRST LARGE MAMMALS OCCUR; PLANET BECOMES DRIER; SMALL EXTINCTION EVENT
		Palaeocene	65,000,000	MANY MODERN MAMMAL FAMILIES EVOLVE; JUNGLES COVER THE EARTH
Mesozoic	Cretaceous	Late	99,000,000	MAMMALS AND BIRDS DIVERSIFY; PTEROSAURS DECLINE; DINOSAURS BECOME EXTINCT
		Early	144,000,000	
	Jurassic	Late	159,000,000	OLDEST MAMMALS, DINOSAURS DOMINATE; FLOWERING PLANTS AND BIRDS; GIANT MARINE REPTILES IN SEA
		Middle	180,000,000	
		Early	206,000,000	
	Triassic	Late	227,000,000	OLDEST DINOSAURS, CROCODILIANS AND PTEROSAURS OCCUR; CONIFERS AND CYCADS BECOME COMMON; EXTINCTION EVENT
		Middle	242,000,000	
		Early	248,000,000	
Palaeozoic	Permian	Late	256,000,000	REPTILES BECOME LARGE; DESERTS PROLIFERATE; GREAT EXTINCTION EVENT DEVASTATES LIFE
		Early	290,000,000	
	Carboniferous	Late	323,000,000	COMPLEX LAND COMMUNITIES OCCUR; COAL FORESTS DOMINATE; OLDEST REPTILES AND INSECTS OCCUR
		Early	354,000,000	
	Devonian	Late	370,000,000	OLDEST SHARKS AND JAWED FISH; OLDEST AMPHIBIANS OCCUR; LOBE- AND RAY-FINNED FISH EVOLVE
		Middle	391,000,000	
		Early	417,000,000	
	Silurian	Late	423,000,000	OLDEST LAND PLANTS AND ANIMALS; CORAL REEFS ARE COMMON
		Early	443,000,000	
	Ordovician	Late	458,000,000	COMPLEX MARINE COMMUNITIES OCCUR; ARTHROPODS DOMINATE THE SEA; JAWLESS FISH BECOME COMMON
		Middle	470,000,000	
		Early	490,000,000	
	Cambrian		543,000,000	THE CAMBRIAN EXPLOSION; OLDEST HARD-SHELLED FOSSILS; OLDEST JAWLESS FISH
Precambrian	Proterozoic Eon		2,500,000,000	THE FIRST SINGLE-CELLED ORGANISMS EVOLVE; THE ATMOSPHERE BECOMES RICH IN OXYGEN; SNOWBALL EARTH; THE FIRST ANIMAL LIFE EVOLVES
	Archaean Eon		4,600,000,000	EARTH FORMS; ITS SURFACE IS BOMBARDED BY METEORITES; LIFE CANNOT SURVIVE

Tree of Life

The following diagrams are a simplified 'family tree' for all the animals portrayed in this book. They show when each creature was alive and also how they are related to one other.

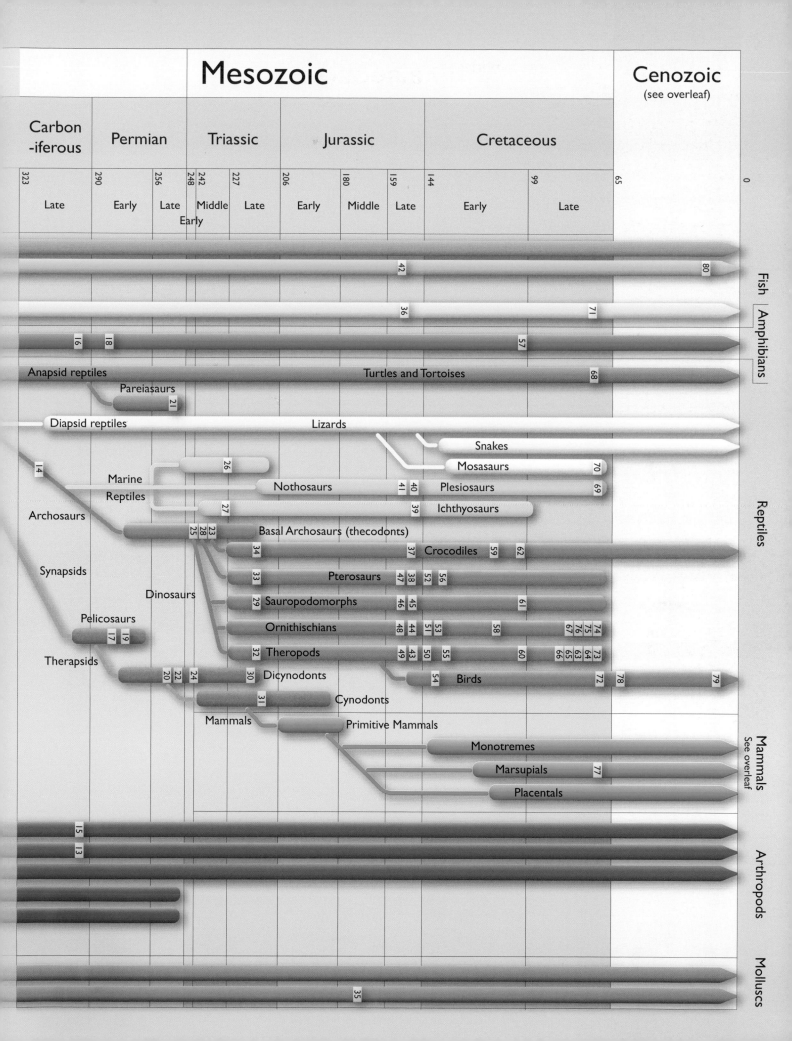

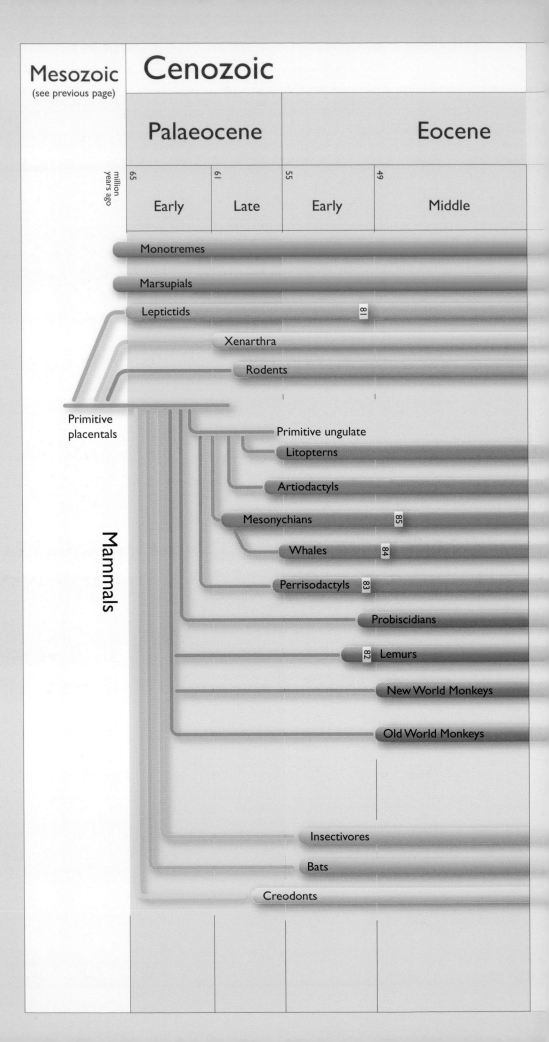

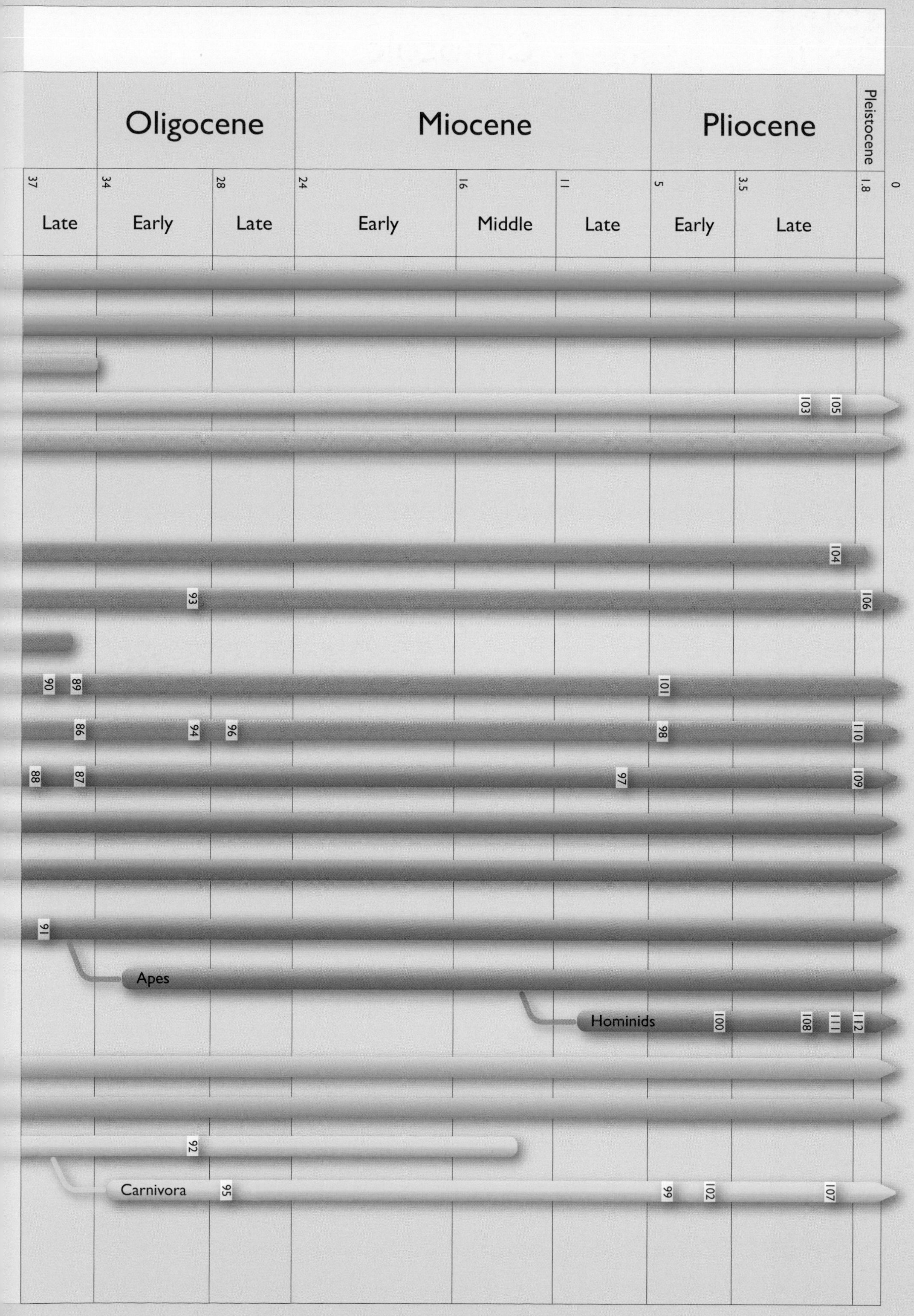

Oligocene
Miocene
Pliocene
Pleistocene
37
34
28
24
16
11
5
3.5
1.8
0
Late
Early
Late
Early
Middle
Late
Early
Late
103
105
104
93
106
90
89
101
86
94
96
98
110
88
87
97
109
91
Apes
Hominids
100
108
111
112
92
Carnivora
95
99
102
107

Acknowledgements

This encyclopedia has been an enormous project, and it would not have been possible without the help and talents of a huge team of people, but from the beginning three individuals have had especially important roles: Mike Milne, who has led the animation teams at Framestore; Jasper James, who has been a key producer on the *Walking with…* series; and Daren Horley, who, as a skin designer and compositor, has been responsible for creating most of the stills.

We would also like to thank the many hundreds of scientists whose work lies at the heart of this project; it was their willingness to share their information, opinions and theories that made this book and the TV series possible.

In addition, we would like to thank Dr Alex Freeman and Dr Jo Wright for their invaluable contribution to the series' research. Thanks also to the staff at the BBC, Discovery, ProSieben and BBC Worldwide, whose constant support and funding helped see this project through to its natural conclusion.

Paul Chambers would like to thank his agent, Sugra Zaman of Watson, Little, Ltd., and his nephews James and Oliver Chambers for their unbounded enthusiasm for prehistoric animals. However, his greatest appreciation is to his wife Rachel, whose ceaseless support and encouragement are a source of inspiration and energy.

Tim Haines would like to thank his family – Olivia, Eleanor, Angus, Rufus and his patient wife Clare – for their support throughout the last ten years.

Picture credits

BBC Worldwide would like to thank the following for permission to reproduce copyright material. While every effort has been made to trace and acknowledge all copyright holders, we would like to apologize for any errors or omissions.

American Museum of Natural History 70 above; Ardea 13 below, 19 above, 97 above, 137, 151 below; Paul Chambers 122; Junyuan Chen 16; Dr Jennifer Clack 30 below; Cleveland Museum of Natural History 26 left; Mike Everhart 132 above; The Field Museum, Chicago 40 left; Scott E Foss, John Day Fossil Beds National Monument, Oregon 172 below; Geo Science Features Picture Library 77; George C Page Museum, Los Angeles/Ed Ikuta 188; Hulton Getty 10; Hunterian Museum and Art Gallery, University of Glasgow 28; Institut und Museum für Geologie und Palaontologie der Universität Tübingen 87; Institut Royale des Sciences Naturelles de Belgique, Brussels 103 above; Institute of Palaeobiology, Warsaw 127; Instituto Portugues de Arqueologia, Lisbon 198; Iziko South African Museum 44 below, 46 left, 48, 69 above; Jasper James 184; Dr D M Martill 104 below; National Museums and Galleries of Wales 38 above; Natural History Museum, London 82 above, 85 below, 100, 142 below, 164, 193 left; NHPA 128; Oxford Scientific Films 13 above, 54; Mike Pitts 18 below, 23, 64, 130; Peter Schouten 206; Science Photo Library 4 above, 18 above, 62, 76, 183; Seapics.com 11; Senckenberg, Messel Research Department 156; Spurlock Museum, Illinois 152; Sternberg Museum of Natural History 134 below; Still Pictures 151 above; Dr Hartmut Thieme, Niedersachsisches Landesverwattungsamt, Institüt für Denkmalpflege, Hanover 204; University of Leeds School of Biology 193 right; University of Michigan Exhibit Museum 166 below; Professor Pat Vickers-Rich, Monash University, Melbourne and Thomas H Rich, Museum of Victoria, Melbourne 112 right; Yale University 130 below.

All other images © BBC Worldwide Ltd or © BBC. Digital images created by Framestore CFC.

Index